[illegible] & [illegible]
Athens, January, 1999

AF575017

Interpreting the Hierarchy of Nature

From Systematic Patterns to Evolutionary Process Theories

Edited by

Lance Grande and Olivier Rieppel

Academic Press

San Diego New York Boston London Sydney Tokyo Toronto

Cover illustration courtesy of Clara Richardson Simpson and used with the permission of The Field Museum, Chicago, Illinois.

This book is printed on acid-free paper. ♾

Copyright © 1994 by ACADEMIC PRESS, INC.

All Rights Reserved.

No part of this publication may be reproduced or transmitted in any form or by any means, electronic or mechanical, including photocopy, recording, or any information storage and retrieval system, without permission in writing from the publisher.

Academic Press, Inc.

525 B Street, Suite 1900, San Diego, California 92101-4495

United Kingdom Edition published by

Academic Press Limited

24–28 Oval Road, London NW1 7DX

Interpreting the hierarchy of nature : from systematic patterns to evolutionary process theories / edited by Lance Grande and Olivier Rieppel.

p. cm.

Includes index.

ISBN 0-12-295120-4

1. Evolution (Biology) 2. Homology (Biology) 3. Phylogeny. 4. Biology--Classification. I. Grande, Lance. II. Rieppel, Olivier.

QH366.2.T675 1994

575--dc20 94-2705

CIP

PRINTED IN THE UNITED STATES OF AMERICA

94 95 96 97 98 99 BB 9 8 7 6 5 4 3 2 1

Contents

Preface

THE ISSUES OF PATTERN AND PROCESS (as defined in Chapter 1, page 1) embody some of the most fundamental concepts in evolutionary biology. But the relative importance of each of these issues, and the degree of their independence from each other is often controversial. This book combines in one volume a diverse assemblage of authors and perspectives, highlighting some of the controversy and breadth that currently exist in macroevolutionary studies. We could have chosen only authors with viewpoints and/or fields of specialty nearly identical to our own for this volume, but we instead tried to incorporate a broader selection of perspectives. It is important, periodically, to shake up our deeply embedded consensus opinions in evolutionary biology and reevaluate that which may have become dogmatic over time. Scientists are expected to periodically question even the most fundamental hypotheses or beliefs in their fields of study. That is what separates science from religion.

It is too common today for biological subdisciplines (e.g., developmental biology, ecology, vertebrate paleontology, various branches of invertebrate paleontology) to have become intellectually divergent and isolated from each other to the degree that when researchers within these different subdisciplines discuss general concepts with each other, such as "homology," "species" or even "evolution," the terminology no longer has a consistent meaning. Comparing fields like phylogenetic systematics to ecology or journals like *Systematic Biology* to *Evolution* illustrates the enormous differences in perspective that currently exist. It is our opinion that if we want to examine the most basic issues of evolutionary biology, we need to identify elements of importance common to all fields of evolutionary biology. We need to identify which controversies are due to semantics, and which are truly contrasting ideas. This is not an easy task, and we do not claim to have come very far in doing so here. But we recognize and draw attention to the problem.

Chapters 2 through 8 in this volume were each written by a different scientist whose research is relevant to modern studies in evolutionary biology. These chapters originated from a series of invited presentations at Field Museum in 1992 on "Systematics and Process" organized by O. Rieppel. Each chapter represents a different specialized approach to the general problems of systematic pattern and/or evolutionary process. Even this seemingly diverse selection represents only a small sample of the diversity of perspectives and approaches that exist today in systematic and evolutionary biology. Chapters 1 and 9 were written by us to comment on some of the more general issues of macroevolutionary studies, to tie together some of the concepts discussed in Chapters 2 through 8, and to present our own views on certain issues (particularly where they might differ from those of other authors in this volume). We include a glossary for the more general reader, which suggests sources for further reading.

We are grateful to those tolerant authors who allowed us to suggest modifications to their contributions in our endeavor to integrate them into a single published volume. In no small part, the success of our efforts is due to the authors' willingness to share their individual perspectives while standardizing their terminology to a degree they may not normally have agreed to. Attempting to standardize terminology for such a diverse group of authors is no easy task in the turbulent world of systematics and evolutionary biology. We hope to have helped the general reader to better identify and appreciate some of the common and the individualistic elements of philosophy and methodology that exist in contemporary evolutionary studies today.

We thank Ms. Elaine Zeiger for helping to put this volume in camera-ready copy, and Ms. Clara R. Simpson for letting us use her original art work on the cover of this volume.

Lance Grande and Olivier Rieppel
Field Museum of Natural History
Chicago

It seemed that the next minute they would discover a solution. Yet it was clear to both of them that the end was still far, far off, and that the hardest and most complicated part was only just beginning.

Anton Chekhov
The Lady with the Dog

1

Introduction to Pattern and Process Perspectives

Lance Grande and Olivier Rieppel

Department of Geology
Field Museum of Natural History
Chicago, Illinois 60605

In the study of evolutionary biology (particularly macroevolutionary studies), two general components can be distinguished: pattern and process. In this context, pattern is the descriptive order of nature (Eldredge and Cracraft, 1980). Process is the mechanism proposed to explain a pattern or order in nature. Two basic types of patterns described in this volume that are relevant to discussions of evolution are hierarchies and temporal sequence. Hierarchies are parsimonious summaries of descriptive data (e.g., cladograms based on morphological characters, biogeographic characters and/or molecular characters; see Figs. 1 and 2 in Grande, Chapter 4, this volume). Temporal patterns describe linear sequences as they are interpreted from stratigraphic order of datable fossils or estimated by molecular clocks (also dependent on stratigraphic position of datable fossils, in part). Both hierarchical and temporal patterns are of retrospective significance for process explanations. Processes could also be thought of as being of two major types: observed and inferred. Observed processes (microevolutionary phenomena) occur in real (humanly observable) time (e.g., birth, components of growth and development, phenotypic population shifts in response to environmental factors and/or variations in gene frequencies, descent of individuals

Copyright © 1994 by Academic Press, Inc.
All rights of reproduction in any form reserved.

from other individuals); inferred processes are not definitively observable in real time but are instead hypothesized to take place in hypothetical (i.e., geological) time (e.g., anagenesis, natural selection of taxa or "species selection", cladogenesis, or descent and origin of new taxa from individuals). Because inferred process is something determined in retrospect, we probably would have no way of conclusively recognizing it in real time, even if we *were* observing a part of it.

The study of evolution must start with the description of organisms. Each set of descriptive data can be arranged in a variety of potential hierarchies. The pattern most relevant to evolutionary studies is a hierarchy of 3-taxon statements supported by maximum congruence of data. Two organisms or taxa can be only similar or dissimilar. But with a comparison of three organisms or taxa, we arrive at the most fundamental statement of pattern reconstruction: A is more similar (and hence believed to be more closely related) to B than either is to C. Any fully resolved hierarchy based on congruence of characters can be rewritten as a hierarchy of subordinated 3-taxon statements.

Congruent data are those data that fit together with no conflict (e.g., see Fig. 1 in Grande, Chapter 4, this volume). Incongruent data are explained as homoplasy (e.g., parallelism or convergence), and homoplastic similarity between organisms can be explained as analogy rather than homology. The taxonomic hierarchy reflecting the most congruent distribution of characters (i.e., the most parsimonious hierarchy) is the most useful pattern to systematic biology for a number of reasons. First, this pattern is the most efficient method of information storage and retrieval (i.e., it is the least redundant summary of descriptive data) (Farris, 1976). Second, it identifies a pattern on which to base an inferred evolutionary hypothesis that is empirically reproducible and testable. Third, the most parsimonious hierarchy provides an element of consistency to pattern analysis because we know how to analyze for the most parsimonious summary pattern of data. Alternatively, we know of no consistent analytical method for identifying the least parsimonious summary of data or some nonarbitrary point in between the most parsimonious and the least parsimonious summaries of data. Only congruent characters can be treated as phylogenetic homologies. (In an evolutionary context, these characters are similarities that are uniquely derived from a common ancestor.) We consider this the only objective and operational concept of homology relevant to phylogenetic studies. We will come back to homology in this chapter and in Chapter 9.

The present volume addresses one of the most fundamental concepts in macroevolutionary studies: the search for patterns in nature, and their relation to evolutionary process theories. The contributing authors address this issue from broadly different and sometimes contradictory perspectives. Most prominent are controversies surrounding the underlying theory of "pattern cladistics" versus "phylogenetic systematics"; the use of molecular versus morphological data in construction of phylogenetic patterns; the use of taxonomic congruence versus total evidence methods in the identification of congruent patterns; the relation of homology to ontogeny and phylogeny; and the phylogenetic information content of unconventional characters. In the effort to bring the different perspectives of the contributing authors to focus on the central theme of the volume, we have attempted to standardize the terminology throughout the chapters to reduce the inevitable confusion that often results from the debate and discussions of general scientific issues by authors with widely different theoretical and practical backgrounds. The diversity of specialization among chapter authors includes philosophy, ecology, developmental biology, paleontology, biogeography, and molecular biology.

Accordingly, the papers in this volume deal with pattern and/or process in a number of ways. The conflict between what has come to be known as "pattern cladistics" versus "phylogenetic systematics" is discussed by several authors in this volume (e.g., Brady, Beatty, Grande, Fisher). Brady (Chapter 2) uses the historical approach to advocate what has been termed the "pattern cladists" position. This position specifies that the search for systematic pattern must be logically independent from evolutionary process theories in order to avoid empirical emptiness of evolutionary studies (in opposition to the so-called phylogenetic systematists' school). Beatty (Chapter 3) puts the controversy between different schools of systematists into the broader perspective of theoretical pluralism, proposing that the search for a unifying theory or explanation may be a misguided pursuit in evolutionary biology. He argues against the independence of systematics from evolutionary reasoning, proposing instead that the two fields are "reciprocally illuminating". Among other things, evolutionary reasoning makes sense of the fact that "theoretical pluralism" is prevalent in ecology, systematics, and throughout biology. The evolutionary history of life would have to be unrealistically constrained in order for us to find a unitary account of each and every domain of biological phenomena. Thus, evolutionary reasoning calls into question the all-too-often unquestioned assumption that we should be searching for *the* correct account of every domain of phenomena, in systematics and in every other field of biology. Grande (Chapter 4), like Brady, advocates the position that the search for systematic pattern must be, as far as possible, logically independent from evolutionary theory. He looks for repeating patterns of order in nature (based on congruent hierarchies of putatively independent sets of descriptive data) as a possible course leading to the documentation of predictable phenomena in evolutionary studies. By defending taxonomic congruence methods in the search for order in nature, he discusses the relative merits of using both total evidence and taxonomic congruence approaches together in the identification of systematic patterns. Fisher (Chapter 6) brings temporal order (stratigraphic distribution of fossils) to bear as arbiter on the choice among equally or nearly equally parsimonious cladograms based on the congruence of characters alone.

Some of the chapters deal primarily with pattern reconstruction. Novacek (Chapter 5) takes a look at the use of morphological versus molecular data in phylogenetic analysis, and the relative merits of these two kinds of data as they are used independently or in combination. Wake (Chapter 7) investigates "unconventional characters" (e.g., sperm morphology) and their potential for testing cladograms based on more traditional kinds of data. Again the question has to be addressed whether these different kinds of data should be used independently or in combination. In either case, "unconventional characters" frequently are merely insufficiently studied characters (in terms of development, individual variation, etc.). Because they are insufficiently known, they may not be completely comparable to more commonly used characters. At one point, for example, molecular data were viewed as "unconventional" but now have become less so, because of continued activity in molecular phylogenetics.

Pre-existing phylogenetic patterns can be used to constrain evolutionary process studies of complex morphological systems. Wake (Chapter 7) critically addresses the protocol now followed by many functional anatomists, who study the evolution of functional morphological systems by mapping these on a cladogram based on other characters. Shubin (Chapter 8) uses cladograms based on independent evidence to investigate the evolution of ontogenies. He also raises the question whether more than one notion of homology can be maintained, and how these would relate to one another. Both these latter authors contrast the phylogenetic concept of homology with a so-called "biological concept of homology" believed to be rooted in shared developmental programs and hence identifiable independent of congruence.

In a final summary chapter of this volume, the coeditors critically review areas of consensus and controversy among the various contributors and a few select issues of particular importance to us. We attempt to synthesize, within a phylogenetic framework, the theoretical background and methodological approaches to pattern/process studies.

References

Eldredge, N., and J. Cracraft. 1980. *Phylogenetic Patterns and the Evolutionary Process.* New York: Columbia University Press. 349 pp.

Farris, J. S. 1976. On the phenetic approach to vertebrate classification. In M. K. Hecht, P. C. Goody and B. M. Hecht (Eds.), *Major Patterns in Vertebrate Evolution.* New York and London: Plenum Press.

2

Pattern Description, Process Explanation, and the History of Morphological Sciences

Ronald H. Brady

Philosophy
Ramapo College
Mahwah, New Jersey 07430-1680

Abstract. Present morphological thought suffers from an unhealthy eclipse of its own history. The particular forgetfulness that afflicts the field may begin with the claims of Thomas Huxley that morphology had been revolutionized by Darwinian theory. Since morphology is a descriptive effort, and Darwinian theory an explanatory one, Huxley's argument appears to confuse two different things. The argument became so strained, in fact, that he was obliged to distort the history of the science to make it fit his thesis. The same confusion and distortion are apparent today, and become particularly apparent in the resistance to "pattern cladistics". The actual history of morphological description is still the best evidence that Darwin's explanation was not the basis of any major accomplishment, and thus that morphological description must be independent of explanatory theory. The independence that history argues, however, can be maintained in present thought only by getting rid of the confusion. For this reason the conceptual structure—that is, the theory—of morphological description must be distinguished from explanatory theory, and the "Darwinian revolution in morphology" dispelled.

Copyright © 1994 by Academic Press, Inc.
All rights of reproduction in any form reserved.

IN AN 1880 ADDRESS BEFORE THE ROYAL SOCIETY titled "The Coming of Age of the Origin of Species", Thomas Huxley spoke of the success of Darwinism since the publication of the *Origin* in 1859. He pictured an earlier geology based on catastrophic revolutions and a view of the fossil record that accepted, in conjunction with these revolutions, large gaps between recognized groups. The *Origin*, according to his account, brought about an abrupt revision of these views, banishing catastrophes in favor of gradualism and finding continuity in the fossil record. Indeed, he continued, the thrust of research since 1859, including his own work on the continuity between dinosaurs and birds, was the closure of fossil gaps. The success of this discovery of continuity was so great, he concluded, that "if the doctrine of Evolution had not existed Paleontologists must have invented it".

Whig history had been written before, and has been written since, but Huxley's lecture seems an unusually good example of the genre, by which history is made to lead inexorably to the lecturer's position. History had led to Darwinism by the time the lecture was given; Huxley's problem was how to make that result seem inevitable. For this reason Huxley first made an exhaustive dichotomy between catastrophe and gaps on one side and Darwinian gradualism and continuity on the other, and then pictured it as decisively settled by the demand, *made by the facts themselves*, for continuity. It was a good rhetorical strategy, and probably impressed all in Huxley's audience who did not remember a different sequence of events.

In his examination of this lecture, Desmond (1982, pp. 58-59) pointed out that catastrophism had waned before the fifties, and that Owen had espoused a gradualistic reading of the record in *British Fossil Mammals*, which he published in 1846, and for which Lyell praised him lavishly. Although Huxley referred to his own work on the gap between birds and dinosaurs, it could have been modeled on Owen's work on the gap between fish and reptiles. Both used a transition fossil–*Archaeopteryx* and *Archegosaurus*, respectively–and argued that it represented a bridge between the two groups in question because it possessed traits of both. But Owen came to this conclusion by 1858–too early for Huxley's paradigm shift–and the continuity of the fossil record discovered by Owen did not call, at least to him, for Darwin's theory.

This last point is potentially the most damaging to Huxley's account. He was at pains to make a necessary connection between Darwinian evolution and morphological continuity. In an 1868 lecture he had claimed that only the Darwinian impulse could ever lead a researcher to attempt to bridge the gaps at all. In the 1880 address he turned the argument about and suggested that after the Darwinian program had demonstrated that gaps were only apparent, this evidence of continuity made the truth of Darwinism visible. Thus not only did Darwinism spur discovery of new facts, it was proven by these same facts. It follows from these claims that Owen should neither have been able to demonstrate continuity nor, having done so, to restrain from embracing Darwin. Yet he did both.

Owen was building what Darwinians would call–or interpret as–a tree, and he was doing so without help from a notion of genealogy. According to Huxley's review of his biological work (1894), Owen made solid morphological contributions but his causal speculations were rooted in *Naturphilosophie* and NeoPlatonism. Since he presumed Owen innocent of any leaning toward Darwinian concepts, how might Huxley have explained why Owen mounted the right research program without the concepts that were necessary to it?

Only with embarrassment, it would seem, but he never had to face the question. Owen, on the other hand, seemed destined to be continually disappointed in his attempts to bring up that very problem. He argued on several occasions that he had announced the continuity of groups before publication of the *Origin*, and that he shared with Darwinians their belief in the progressive appearance of new species in the fossil record. Owen thought that this continuity was an empirical discovery which called for explanatory hypotheses, and was bewildered by claims that the evidence was the private possession of Darwinian explanation. He wrote, in 1860 (p. 403),

> As to the successive appearance of new species in the course of geological time, it is first requisite to avoid the common mistake of confounding the propositions, of species being the result of a

> continuously operating secondary cause, and the mode of operation of that cause.

Desmond (1982, p. 63) found a letter from the Duke of Argyll that reassured Owen on this very point:

> You are quite justified in drawing a broad distinction between "Creation by Law"–(*some* Law)–and the special theory in respect to the precise nature of that Law, which Darwin has put forward.

In other words, we may distinguish an empirical regularity, which calls for a cause, from a particular causal theory brought forward to explain the regularity.

We may suppose, therefore, that whatever the opposition between Owen and the Darwinians, their discovery procedures evidently shared important elements. These shared elements, however, were not part of the historical explanation, for on that point there was no sharing, and therefore they must be found in the procedures by which pattern is *detected* and *described*. Huxley collapses the distinction between description and explanation, as much evolutionary rhetoric has done since, for this allows him to make Darwin the revolutionary in morphology. Unfortunately, the result of this collapse is merely confusion, whether it is performed in the 1800s or today. The rest of this chapter will examine the role of description in the history of morphology and in current arguments. So little is said about description in taxonomic literature, however, that before I continue it seems prudent to examine this phase of science in greater depth.

THE THEORETICAL COMPONENT OF DESCRIPTION

Observation, as N. R. Hanson argued in 1958, is theory laden, and his statement will need a gloss before we can proceed. Hanson's conclusion does not reduce the usefulness of observation to science, but rather announces the demise of a naïveté that assumed that interpretation begins only *after* we have gained observations to interpret. This was the position of the sense-data theorists, who supposed that perception was a mechanical transmission of the senses, free from conceptual activity. Against this, Hanson argued that we must recognize two forms of interpretation–how we interpret what we see, and how we interpret *in order to see.* Thus observation is already a product of interpretation, for the human perceptual process cannot be innocent of conceptual elaboration.

The fact that concepts must be included in perception was discovered for the modern world by Kant, but it has antecedents that stretch back though medieval thought into antiquity. However one cares to argue the point, all accounts have the dichotomy between the sensible and the conceptual in common. The heart of the argument is that while the operations of our sense organs present us with sensations without conscious effort on our part, all *relations* are conceptual, and must be added by the intentional activity of the observer. Because nothing is intelligible without relations, whether internal or external or both, observation is the result of conceptual elaboration of the sensible report (which report would indeed be useless, because unintelligible, without conceptual elaboration).

Observation is therefore a product of "theory", because it is a product of conceptual elaboration, but this admission does not, in itself, destroy the validity of empirical results. Neither circularity of description or of explanation is a necessary result.

The suspicion of circularity of description rests on the assumption that our thinking *arbitrarily* adds the relations it proposes to the sensible report. It is obvious, however, that we cannot perceive whatever we like, but only what the sensible report allows us to perceive. The conceptual faculty only *proposes* relations, which proposal must then produce a perception of the same relations (now immanent in sensible structure) if it is to be successful. If conceptual proposals are not successful our thinking moves on to something better, but the replacement often happens too quickly to be noticed. The "double-take" is the rare occurrence that *is* noticed, for in this case we begin to form a perception which, having resulted in contradiction on some level, is dissolved again and replaced by the second "take". If thinking must propose the relation in the first place, the sensible report determines which proposal is successful, or more successful.

Success consists in seeing intelligible structure. Because we do this whenever we open our eyes we are apt to forget the intellectual activity involved. But the moment our habitual mode of perception is frustrated, by camouflage let us say, the conceptual work of perception becomes conscious. Camouflaged figures often become visible when we are given a sketch of the shapes to look for, where before that revelation their camouflage made them impossible to see by denying some sensible clues that would normally mark a border. Once we look for the right shapes, however, camouflage ceases to mislead. *All* seeing is led by concepts, even as our penetration of camouflage was, but when habitual expectations are met we no longer notice the fact.

Once we *have* seen we may arrange what we have seen in logical schemes–for example, we arrange perceived forms in a hierarchy–but this categorical elaboration merely makes explicit the relations implicit in the perceptual act. The elaboration puts a new face on the same data–like the difference between seeing a group of trees and seeing that there are five of them. We have not added anything to the trees by numeration. Even so, we do not add anything to the topological variants by arranging them hierarchically. If hierarchical relations–various levels of sameness and difference–were not already part of the observed structures they could not be abstracted to form a hierarchy. (I shall take up the problems of completeness and inconsistency, which are other matters, in a later section.)

Circularity of explanation, on the other hand, remains a serious difficulty. If a set of observations were found to be a *product* of our explanatory theory the same observations could not be used to *test* the theory. When we consider that the lack of testing power would mean that observation did not constitute evidence for theory, the complaint is deadly. But how could observation be an artifact of theory? By the simple expedient *describing* phenomena in terms of explanation–that is, using the language of explanation to describe relations in the data. Results derived from a fixed explanatory framework build explanation into observation, producing that most desired of outcomes–experience that explains itself. As I understand it, the notion that texts read themselves, whether we mean by "text" a set of writings or a set of phenomena, is the strategy of fundamentalism, which does simplify the world, but also makes science unnecessary.

But of course phenomena do not interpret themselves, and the fundamental concepts necessary to elaborate relations in taxonomic data are logical rather than causal, as any cursory examination of the history of taxonomy will show. The fact that the theoretical component of our observations *may* be drawn from our explanatory theory does not mean that it must be. Our descriptive concepts–those used to formulate perception–need not be the same as our explanatory concepts. In this case the circularity that destroys the notion of evidence would be avoided. This principle seems sound, and that is why it is surprising to find some authors invoking explanatory concepts in the descriptive phase, a practice that gives rise to a debilitating circularity.

Huxley walked over that line when he attempted to make Darwinian explanation necessary to discovery of patterns it was invoked to explain. He says this, of course, in order to claim success and evidence for his program. Success, in that it made a unique prediction and then confirmed the same. Evidence, in that this confirmation could be explained only by Darwinian theory. The first claim is contradicted by the existence of Owen's work, the second, by reason. When the accomplishments of the period are reviewed, one can agree with both Owen and Huxley that the closure of the fossil gaps is a major project, and with Owen that however the pattern is explained, both sides agree on the progressive appearance of life–read "differentiation of major groups", and the morphological continuity between groups. The principles held in common, therefore, are the *descriptive* concepts necessary to elaborate differentiation and continuity.

CLOSING THE GAPS IN 1859

Let us re-examine the issues set forth by Huxley in the light of the preceding arguments. The "gaps" in question opened between major groups, like fishes and tetrapods, or birds and everything else. If one held that these groups were entirely discrete, the gap–that is, a discontinuity in morphology–was to be expected. The revolution in morphology that Huxley describes consisted of a proposal of continuity that led to a detection of the same. This sequence seems to be modeled upon actual events, but events that had already taken place by the publication of the *Origin*. As two recent studies (Appel, 1987; Laurent, 1987) have shown, we must look back to the Paris debate between Geoffroy and Baron Cuvier, and its aftermath, to find the transition after which Huxley constructs his fiction.

The battle lines of that famous event are indeed those of discreteness and continuity. Cuvier and Geoffroy had been long-time friends, although their approach to biological form was divergent. By the time of the debate (1830) Cuvier was recognized as the greatest anatomist in France, and his influence was so commanding that to some he was the establishment, rather than its best representative. He had opposed Lamarckian evolution, held for the fixity of species, and demonstrated that the varied forms of animals could be *teleologically* understood–that is, that structure reflected function. He had looked on with interest as Geoffroy developed his "philosophical anatomy", tracing "affinities" of organisms on the basis of "analogous" organs, an idea generally recognized but not previously given a central position. He gave Geoffroy credit for working through homologies, as they are now termed, of vertebrate structure to a greater degree than had previously been attempted, but he noted that this mode of comparison, while as old as Aristotle, was but a secondary principle. The primary form of comparison was teleological.

If teleology is the basis of comparison of organisms, the regularity discovered by observation should be the covariance of structure and function. Different groups should possess unique structures, commensurate with their lifestyles, and where homologies between these groups exist so does continuity of function. Geoffroy's proposal of complete morphological continuity between major groups represented, to Cuvier, an excess doomed to failure. In 1828 he stated unequivocally that fishes were a discrete group with unique structure.

In order to provide a morphological basis of comparison, Geoffroy made the first attempt at rigorous definition of what now appears to be a familiar standard–the "principle of connections" –postulating that within the common plan of a group a standard part, or "analogue", was identifiable by its *connections* to other parts–that is, analogues occupy the same position in a topological schema, or can themselves be mapped, according to their internal composition, on the same schema. In his *Philosophie anatomique* (1818) he argued: "we will always find in each family all the organic materials found in another", for "an organ is sooner altered, atrophied, or annihilated, than transposed". The "organic materials" are the units recognized by connections, and the real nature of the "common plan" is a set of connections, a kind of map, within which the elements connected may alter in shape, size, orientation and function, but not in the order of connections–throughout all changes the positional schema is conserved. Terminal organs may even be subtracted (annihilated), if there is no use for them, but subtraction is not transposition or creation of new materials.

Geoffroy did not refrain from speculation on the cause of this constant topology. He hypothesized a mechanism based on blood-flow, and set himself in opposition to notions of divine intervention or *vital force* in favor of naturalistic explanations. He spent most of his effort arguing that animals do indeed share a constant topology, however, and when he did advance causal speculations he justified them by the need to explain his topological results. His explanations do not seem to have had much impact on succeeding generations, but the opposite is true of his descriptive proposal.

Disagreement between Cuvier and Geoffroy broke into open dispute before the French Academy in 1830. While both Cuvier and Geoffroy presented extreme positions in the debate (Geoffroy suggested that even *embranchements* were transforms of one another), the underlying question was clear. Would groups be looked upon as teleologically distinct, or morphologically continuous? After the debate both sides claimed victory, and as the dust settled biologists began the task of integration. The future of the science would consist of some combination of teleological and morphological approaches.

Let us remember that Geoffroy's proposal concerned the "ideal"–read "logical"–relations of form. It rested on the postulate of a set of invariant topological relations–termed "connections"–within a variation of potentially everything else. Forms compared by this standard become differing executions of the *same* topology, and since they are isomorphic on this level, each must relate to any other as a *transformation* of that other. I must emphasize, however, that in this context the term "transformation" indicates pattern and not process. Thus Owen defined "homology" (a word with topological significance in geometry) as "the same organ under every form and function", where sameness is recognized by identity of connections. Gaps between major groups are closed by understanding those groups to be continuous through transformation (differing executions of invariant topology), and "bridge" fossils provide a "graded series" to clarify homologies. Again, the transformation postulated here is topological only, and thus is a *logical* rather than *historical* relation. This same relation, however, was to provide the evidence used to support hypotheses of physical transformation when such hypotheses were advanced.

By the time Darwin published, he looked on some aspects of Geoffroy's work as authoritative, and made appeal to them in Chapter 13 of the *Origin*: "Morphology" (1859, p. 434).

> What could be more curious than that the hand of a man, formed for grasping, that of a mole for digging, the leg of the horse, the paddle of the porpoise, and the wing of the bat, should all be constructed on the same pattern, and should include the same bones, in the same relative positions? Geoffroy St. Hilaire has insisted strongly on the high importance of relative connection in homologous organs: the parts may change to almost any extent in form and size, and yet they always remain together in the same order. We never find, for instance, the bones of the arm and forearm, or of the thigh and leg, transposed. Hence the same names can be given to the homologous bones in widely different animals....

Darwin's next paragraph invoked the revolution of which I spoke above, for it correctly opposed Cuvier's teleology to Geoffroy's morphology:

> Nothing can be more hopeless than to attempt to explain this similarity of pattern in members of the same class, by utility or by the doctrine of final causes. The hopelessness of the attempt has been expressly admitted by Owen in his most interesting work of the *Nature of Limbs*. On the ordinary view of the independent creation of each being, we can

> only say that so it is;—that it has so pleased the creator to construct each animal and plant.

Notice that Darwin appealed to the authority of Owen in rejecting teleology. He acknowledged this debt because Owen, in *On the Archetype and Homologies of the Vertebrate Skeleton* (1846) and *On the Nature of Limbs* (1849), had worked through the issues of the Cuvier-Geoffroy debate and placed Geoffroy in a secure position. (In these works Owen substituted "homology" for commonality of structure, and used "analogy" for commonality of function.) In order to secure Geoffroy's morphology, however, Owen had to answer the teleological argument.

Cuvier's teleology postulates that structure is determined by function, leading to a discontinuity of structure whenever there was a discontinuity of function. Owen answered that this view, like that of the protestant theologians in Britain (one should call to mind Paley's argument on the watch), would design animals to fit their mode of life as machines are designed to fit their function. But in machines, he points out, both the materials and the *plan* of construction are determined by the function. Even Cuvier would have to admit, however, that animals often adapt a single underlying plan to several functions. If the same structure supports differing functions, topology is more basic than teleology, and the admission of such topological invariance reveals the failure of the teleological program. "The teleologist", wrote Owen (1849), "would rather expect to find the same direct and purposive adaptation of the limb to its office as in a machine", but morphological investigation shows that limbs built for flying, running, swimming, digging, for example, are all constructed on the same plan.

Darwin's gloss of Owen continued with Owen's modification of Geoffroy's postulate that all vertebrates contain the same bones, or at least, the same centers of ossification. Finding that no juggling of skeletons could produce the same "organic materials", or even the same centers of ossification, Owen tried another strategy. He generalized on all skeletal forms by finding a common denominator: the vertebrate archetype. This structure, essentially a repetition of vertebrae, would serve as a basis if one preserved enough teleology to modify the vertebra as the need demanded. Thus the skull area of the archetype consists of four vertebrae, which becomes the modern skull by an expansion of the spinal cord and a consequent ballooning of the vertebrae. By extension and metamorphosis, vertebral processes become ribs and sternum, and finally, becoming "teleologically compound", such extensions form limbs. The schema is too arbitrary to produce empirical confirmations, but that problem was not immediately apparent to Owen or his followers.

The archetype itself, when cast in diagrammatic form, looks like a fish. To the literal-minded Darwin, that is exactly what it signified. Thus he postulated an "ancient progenitor, the archetype as it may be called", and filled out the rest of the program after Owen's ideal schema, approximating a teleology of design through his own "natural selection" (1859, pp. 437-438):

> An indefinite repetition of the same part or organ is the common characteristic (as Owen has observed) of all low or little modified forms; therefore we may readily believe that the unknown progenitor of the vertebrata possessed many vertebrae; the unknown progenitor of the articulata, many segments; and the unknown progenitor of the flowering plants, many whorls of leaves. We have formerly seen that parts many times repeated are eminently liable to vary in number and structure; consequently it is quite probable that natural selection, during a long continued course of modification, should have seized on a certain number of primordially similar elements, many times repeated, and have adapted them to diverse purposes.

As we have seen above, Darwin moved away from transcendental entities. Geoffroy's school of "philosophic anatomy" dealt with the "ideal" or "philosophic" relations of the forms, but Darwin wanted to arrive at something more concrete. In his chapter on "Doubtful Species" he had mentioned that some forms are so closely linked by intermediate gradations that naturalists are unsure of how to rank them. He then points out that when viewing such a series of graded forms one seems to be seeing a *process* of change (1859, p. 51):

> These differences blend into each other in an insensible series; and a series impresses the mind with the idea of an actual passage.

The notion that historical process is immediately seen in morphological evidence runs through the language of Darwin's text, culminating in his advice that a "philosophic" approach to the evidence is unnecessary (1859, pp. 438-439):

> Naturalists frequently speak of the skull as formed from metamorphosed vertebrae; the jaws of crabs as metamorphosed legs, the stamens and pistils of flowers as metamorphosed leaves.... Naturalists, however, use such language only in a metaphorical sense; they are far from meaning that during a long course of descent, primordial organs of any kind—vertebrae in the one case and legs in the other—have actually been converted into skulls or jaws. Yet so strong is the appearance of a modification of this nature having occurred, that naturalists can hardly avoid employing language having this plain signification. On my view these terms may be used literally....

The metaphoric quality of previous language was produced by the mental reserve that described the forms of a graded series "as if" they pictured an actual passage, or the topological identity of two forms "as if" one was produced by a physical distortion of the other. We might use the same terminology in describing the relation of geometric forms when those forms are produced by a systematic change in the execution of a constant topology. A series of parallelograms, for instance, which begins with a right-angled form and departs successively from that, might be described in terms of physical transformation by suggesting that "We take the right-angled box and gradually squeeze it flat". Such descriptions are obviously metaphorical, aimed at the set of relations by which the forms are related rather than a process by which they are formed. After all, the "passage" or movement from one form to the other, whether in the geometric or biological example, is contained in the human imagination rather than the static shapes. This is why Geoffroy termed it "philosophical" rather than actual.

For Darwin these distinctions must have seemed merely fainthearted. In the quote above, Geoffroy's "philosophical" relations become *merely* figurative ones, the "plain signification" of which, if we will but recognize it, is "an actual passage"–i.e., a process. It would seem that the collapse of the distinction between description and explanation was central to Darwin's argument before Huxley adopted it.

If we take Darwin at his word, this literalism, with its attendant genealogical explanation, is his real contribution to the science of morphology. Beyond this he merely adopted earlier integrations of teleology and morphology, and according to Laurent (1987) he was not even the first to explain logical transformation by descent. But I do not care to address claims of priority here–it is enough that Darwin himself limited his contribution to a literal interpretation, arguing that he could "see" the events of his explanation in the morphology.

The project of "closing the gaps" was neither invented by Darwin nor claimed by him, because he admitted that it had been adopted from Owen, and thus ultimately from Geoffroy. He does take credit for changing the proposal from a "philosophic" to a "literal" one, but the enabling insight was still drawn from Geoffroy's technical advances, and the search for continuity is always performed as a search for topological similarity. Although the evolutionary meaning of "homology" now obscures the original topological concept, it is still recognizable. As Goodwin wrote in 1984, using Owen's term: "Homology is an equivalence relation of a set of terms which share a common structural plan and are thus transformable into one another. This is therefore a logical relation independent of any historical or genealogical relationships which the actual structures may have."

TOPOLOGY AND HIERARCHY

The advance in description that I have been tracing was the attempt, however incomplete in results, to compare on the basis of invariant topology. In order to examine the implications of this principle more closely, however, I turn briefly to the second descriptive proposal that grounds taxonomy.

If animals can be compared on the basis of a common-plan, standard-part order, or what Owen termed "special homology", the results of that comparison should provide a measure of resemblance, which measure has been found in a hierarchy of topological variation. (In order to avoid confusion at this point I should perhaps mention that a proposal of hierarchical ordering of special homologies differentiates this research program from that of phenetics, which compares organisms on the basis of "overall resemblance". Whatever other differences accrue at later points, the main difference between cladistics and phenetics rests not on the hierarchical proposal but on the means of comparison. The former is topological in the sense defined above, the latter is not.)

A hierarchy of less general variants subordinated to more general ones is a logical possibility that the standard of comparison admits. That variations should have a regular distribution, however, or that this regularity should be hierarchical, does not follow from the standard of comparison, and must be determined empirically.

In order to make this determination we must again propose the relations we seek before we can expect to find them, since relations must be thought before they can be perceived. The hierarchical proposal is older than Aristotle, but while not unchallenged, the proposal has lasted through many periods of biology, presumably because it enjoys reasonable empirical success–i.e., congruence between observed data and conceptual relation–whatever our explanatory reasons for liking it. After all, differing proposals of order may be compared on their measure of descriptive success, without reference to explanatory justification. It follows that the proposal could be replaced by any order that could show greater success–greater congruence between data and conceptual pattern.

That descriptive proposals can be justified only by descriptive success is not well understood. Panchen (1992, p. 239), wonders why I decline, in a 1985 paper, "to postulate that the natural order is a comprehensive hierarchy". He clarifies his concern a few pages later (p. 243) when he remarks that "the existence of a hierarchy embracing all living things is in doubt", and adds that unfortunately "the only valid method of revealing it [the hierarchy] is that of cladistics, which depends absolutely on its *a priori* existence. This dilemma is insoluble." I agree that *this* problem is insoluble, which is why I am relieved that cladistics need not make hierarchy into an *a priori* truth. I cannot see why any science worthy of the name should dictate the results of observation before observing. Obviously we "reveal" nothing about nature through *a priori* decisions.

The circularity that Panchen raises is created by describing the data in terms of the explanation–that is, by postulating that "the natural order is a comprehensive hierarchy", or supposing such a postulation necessary. But as I have been arguing, hierarchical order is not part of the explanation, but part of the description, and is produced by logical elaboration of topological data. Since it is an empirical postulate, it is always provisional, and should not be legislated. The empirical success of hierarchical ordering has established this order as the most successful so far, and that is justification enough for a provisional postulate. There will be reason to shift our procedures when and if more successful proposals of relation show up–and as long as we have not succumbed to *a priori* demands, we shall be able to do so.

Although parsimony is a side issue, it should be briefly examined. Having proposed a hierarchical order, we will need (as in the case of topological comparison) methodological principles by which to select, from a plethora of possible hierarchies, the one that serves our purpose. Upon reflection, our choice of order has already determined that method. The attempt to elaborate hierarchical relations in data proposes that the data are part of a specific hierarchy which we desire to recover. If all character information were congruent with a single hierarchy then selection would be automatic. If some characters contradict the report of others, however, we select the most complete hierarchy that can be constructed from the data, which is the hierarchy with fewest contradictions (incongruence) between data and hierarchical order. The hierarchy selected in this manner will be best supported by evidence, for it maximizes congruence between data and hierarchical pattern.

The procedure for this selection is currently termed a parsimony program. If included in the explanatory concepts parsimony would carry ontological implications, but at the descriptive level it does not do so.

Of course, its use implies the methodological assumption that the best supported hierarchy is the best candidate for the original order. This does not seem an unlikely procedure, considering that it is also the only candidate. Any alternative hypothesis throws doubt on the framework proposal that the data record a hierarchy. Due to its methodological status, such a method of minimizing incongruence could be used on any data set, taxonomic, linguistic, or other, in order to find the most complete hierarchy that the set admitted.

With this much understood, we can see also that a further principle is entailed in the decision to use a data set that is not completely congruent. Data are elaborated according to the proposal of hierarchical relations, but this proposal necessitates the assumption that the hierarchy is the real order of correspondences. By this logic, if the hierarchical order is real, incongruence is not. Or rather, *random* incongruence is not real–i.e., it is a product of human error rather than an aspect of nature. (Let me caution here that I do not mean to legislate an *a priori* rule. I am only working through the logic of the postulate, always provisional, by which we propose to order topological data.)

Nonrandom incongruence–incongruence that shows a repeated pattern– represents an order that lies outside the hierarchy of congruent characters. Hybridism, for example, can produce two equally parsimonious hierarchies for the same organism, differing only through the placement of one terminal branch. But random incongruence, because it can be included only by a random set of hierarchies, is not evidence for hierarchical order at all. Recognition of such incongruence as real would seem to reject the postulation on which the whole program is based, since it assumes that nature contradicts the proposal of hierarchy. (Obviously, if incongruence were the *general* result of our attempt to order data, we would have little reason to suppose the hierarchical proposal successful.)

Because random incongruence is explained as error, it provides an added test of "homology". The parsimony program distinguishes those apparent correspondences (homologies) which are distributed hierarchically from those which are not. If *real*, topological correspondence is hierarchically distributed; the postulates of correspondence which produce random incongruence are erroneous. This test of correspondence–by taxic congruence–may appear to take precedence over direct topological comparison because it is able to correct the former, but the impression is illusory. The hierarchy is one of topological variations. Congruence is able to test *individual* postulations of correspondence because it shows them to be irregularities in a larger order. Since that order is produced by combining results of many direct comparisons, it is ultimately topological evidence that is brought to bear on the suspect homologies, and the generality of that evidence is decisive.

Present taxonomy is therefore grounded on two empirical proposals–that organisms may be compared topologically and that the variants revealed by such comparison are ordered hierarchically. Implicit in these proposals is that no other form of description reveals such a strong regularity. Darwin accepted the same procedures in Chapter 13 of the *Origin*, and his explanation was matched to this description. Because their content is descriptive, the same proposals are found in Owen and his French forerunners. Huxley's notion of a Darwinian revolution in morphology is just as untrue when applied to present morphology as it was when applied to the morphology of 1880.

CLOSING THE GAPS TODAY

As the last section has argued, continuity in morphological evidence is found descriptively, if it is found at all. Huxley's version of events did not recognize this demand, and I have been at pains to correct his history. The central problem, however, is Darwin's literalism, which Huxley was applying. It is still a problem, although it is rarely noticed. Since the victory of Darwinian ideas in the realm of opinion the popular view holds, with Darwin (1859, pp. 438-439), that "during a long course of descent, primordial organs...have actually been converted..." to very different structures. Of course, we do not observe the actual process: "Yet so strong is the appearance of a modification of this nature having occurred, that naturalists can hardly avoid employing language having this plain signification." It is obvious that the verbal habit is widely shared today. Darwinian literalism, in fact, controls the *imagination* of the modern to such a degree that the term "transformation" has only the "plain signification" mentioned above–i.e., a temporal process in which the earlier form changes into the later. Very little attention is paid to the circularity that results from describing the morphology with terms drawn from explanatory theory.

Given the prevailing mood of literalism, one of the most interesting comments on morphology in recent years was Patterson's (1982) argument that one finds two distinct meanings of homology in present systematics. The one that he prefers he terms "taxic homology" and gives as an example the more or less inclusive homologies (synapomorphies) that allow our arrangement of the groups within the Gnathostomata. Each of these homologies implies a hypothesis of grouping, and thus can be tested both by topological similarity and by congruence.

On the other hand, the attempt to work though what he calls "transformational homologies"–like the hyomandibula of fishes with the stapes of mammals–advances no new hypothesis of grouping (the postulation depends upon the inclusion of both fishes and mammals in the Gnathostomata) but makes a hypothesis of literal transformation. Patterson argued that such hypotheses produce no empirical confirmations since the topology is unclear and the test of congruence cannot be applied (no new hypothesis of grouping is implied). Patterson labeled this form of homology "vacuous" of empirical content, arguing that only taxic homologies could be submitted to any meaningful test.

As Patterson has demonstrated, transformational homologies do not carry any new hypothesis of grouping, so the generality of evidence cannot be used to test them. That leaves topology, and if this is also indeterminate, transformational homologies will be untestable. The taxic approach applies the descriptive concepts of topological constancy and hierarchical congruence to elaborate the raw data, and its empirical content consists in the same data with its relations made conceptually explicit. Lacking these descriptive tools, transformational approaches reduce to applications of explanatory hypotheses to descriptive projects–that is, the discovery of homologies. Unfortunately, an explanation of observed facts does not add to the facts, however likely the scenario may seem. The approach is empirically empty because it is an exercise in circularity. The relations made explicit are not those of the data but those of the explanation.

How could we deceive ourselves in this manner? In a discussion of the distinction between taxic and transformational, Rieppel (1988, pp. 111-112) presents the transformational approach as an artifact of Darwinian literalization:

> The taxic approach results in a hierarchy of fixed types representing logical relations of form, and in that sense it is ideal as well as static. The transformational approach, however, puts the hierarchy of homologies into a temporal sequence of form in order to ask the question of evolutionary transformation. The static hierarchy of types is reduced to a temporal sequence of forms, homology comes to denote *relations of succession*.

The complaint is not that we account for the phenomena by a hypothesized historical sequence, but that we actually describe them in this manner, seeing "an actual passage" rather than a hierarchy of types in the evidence. Once we have accepted the literal notion of transformation, that is what we look to see in morphology. Thus we try to identify ancestry–i.e., what changed into what–based on morphological evidence, as if a series of triangles contained evidence of parentage. Animals do possess ancestry, while triangles do not, but *what we can determine about either subject must depend upon the nature of the evidence examined.* Since in either case that evidence consists of the comparison of forms by topological standards, a set of topological relations will be the result. In the case of morphology, the set that Rieppel terms the "hierarchy of types".

Darwin's mode of thought shifted the context of morphological arguments to the degree that Huxley could make his remarks with little fear of challenge, then or now. I think that the real effect of Patterson's argument must be to shift the context of discussion once again. His critique of transformational homology makes the logical priority of the taxic homology quite clear, and his finding that transformational homologies are usually empty of empirical content is in keeping with the circularity of describing data with the terms of explanation. Due to the ubiquitous literalism that he addresses, however, he uses the term "transformation" to refer only to literal change, and does not challenge the literalism of language that maintains the literalism of thought he has faulted.

TRANSFORMATION REVISITED

Rieppel (1988) argued that the hierarchy of types was "static". He meant by this that it was innocent of temporal process, or what most writers seem to mean by "transformation". But certainly, when a topologist speaks of one form being a transformation of another no temporality is implied. A transformation in the most general sense is a particular kind of change, one which both preserves and alters–that is, the underlying identity must remain constant while the appearance or execution of that identity changes. By this standard any homology is a transformation, for it is clear that the idea demands both underlying identity and external difference. Homologous organs are not "the same" physically or sensibly, but can be equated by possession of the same relations. Thus a particular tetrapod limb can be considered a topological transform of another, in that it can be geometrically produced from that other by a set of distortions of the parts.

Of course, "transformation" in this sense has no historical, or even directional, meaning. *One limb does not turn into another*, but both are isomorphic products of the same topology. Direction, or polarity between the more general and the less general variants within the group of compared forms, is also a logical relation, but one that enters only with hierarchical ordering. No such ordering is possible, however, until we have a basis for comparison–that is, until we have established that the forms compared *are* transformations (of a constant topology). Thus transformation is established before direction, and is independent of any temporal reference. (This independence cannot be reversed. Theories of temporal transformation need evidence, and only forms exhibiting topological transformation can provide it.)

The basis of all taxonomic projects is the description of topological relation, and this description is itself a major scientific problem. The invariant element underlying transformation is clearly the point at issue, for a successful formulation of that element would put morphology on a rational footing. Geoffroy tried to make such a formulation with his *unity of organic composition*–the notion that each organism could be compared as transformations of the same "materials" or topological elements. If he had been able to demonstrate a total invariance for all vertebrates, the phylum would consist of varied appearances of the same (the same *law* or relation, not the same *thing*), and while the major groups would show discontinuities of execution, they would demonstrate a continuity of plan. But Geoffroy's research program fell short of its goal. The principle seems very clear when we are dealing with tetrapod limbs, but if we try to extend the analysis to sarcopterygian fins ambiguities arise. Geoffroy's own identification of the bones of the gill cover of fishes with those of the inner ear of tetrapods succeeded in showing only that these might be homologous, but it produced no test for this conclusion.

Of course diagnostic homologies for the Gnathostomata exist. But what remains unclear is the implication for the whole group of this partial identity. Given that a set of homologies identify both fishes and tetrapods as gnathostomes, does the *unity of organic materials*, or invariant topology, imply other parts must be homologous as well? Do sarcopterygian fins therefore represent a topological transformations of tetrapod limbs? (I deliberately reverse the probable temporal sequence to emphasize the nature of the question–we are inquiring about topology, not sequence.) We may assert this unity, as the literature of transformational homology can attest, but we may also suppose that both be considered unique and unrelated structures. Either choice makes an interpretation of the principle of invariant topology.

Empirical tests cannot at present choose between these alternatives for it is at this point that the research program–comparison by recognizable invariance–fails. I mean that it tells us neither whether the fin is or is not homologous with the limb. Rieppel (1988, pp. 44-49) has reviewed the limitations of the topological judgment, implying the same conclusion. At present, a decision that fin and limb, or gill and middle ear bones, must be homologous, or need not be, is not determined by evidence, but by our interpretation of the postulate by which we compare organisms–invariant topology. And differing interpretations are still possible, for almost two centuries after Geoffroy made topological comparison the basis of morphology we still cannot determine empirically the degree of invariance that organisms possess.

Our comparisons begin from some form of Geoffroy's postulate, which we assume *in order to see* resemblance, not to interpret the resemblance we have seen. The fact that the degree of invariance cannot be determined empirically in some cases indicates either (1) that the postulate is false, or (2) that our criteria are inadequate to our intention. If the postulate is false our method of comparison may be secondary, as Cuvier claimed, for it would seem that whatever determines when topological invariance will be conserved and when it will not is more important, in just the sense that teleology once seemed more basic. If the postulate is true, our approach is correct (and our groups largely valid). But the research program that could formulate the underlying constant, rendering it recognizable to human perception in all cases, does not yet exist.

How do organisms vary? Our present methods of tracing topology are quite crude, to say the least. We are often guided by intuition, but we do not possess the means to make intuition explicit. Mathematical topology can demonstrate transformations which look quite discrete to the uninitiated–we do not know whether organisms are giving us the same problem. And research into the nature of biological transformation has barely begun. Upon reflection, it is somewhat odd that morphology should have advanced so little toward the solution of a problem that Geoffroy formulated so well. One might even speculate that the Darwinian explanation of taxonomic relations was, in a sense, premature, since the logical nature of those relations is still unclear.

To some degree the success of the hierarchical postulate must have distracted attention from the incomplete formulation of topological invariance. After all, when a hypothesis of topological invariance implies a grouping it can be tested by congruence, and if the generality of evidence (congruence) confirms the hypothesis, our problem seems at an end. But this success does not seem adequate to explain the apparent eclipse of the descriptive problem for generations of taxonomists. Yet judging from the literature something like this has happened. Morphology is rendered possible by the theoretical postulates which elaborate the phenomena and provide a standard of comparison. Those postulates are the "theory" of morphological description, and they are just as capable of investigation as explanatory theories. Unlike the latter, however, they have suffered a period of neglect, probably beginning about 1859.

References

Appel, T. A. 1987. *The Cuvier-Geoffroy Debate*. Oxford, New York: Oxford University Press.

Darwin, C. 1859. *On the Origin of Species*. (Facsimile of the First Edition, New York: Atheneum. 1972.)

Desmond, A. 1982. *Archetypes and Ancestors*. Chicago: University of Chicago Press.

Geoffroy Saint-Hilaire, E. 1818. *Philosophie anatomique: des organs respiratoires sous le rapport de la determination et de l'identité de leurs pieces osseuses*, 98-99. Paris. Reprinted Brussels: Culture et Civilization. 1968. Translation quoted from Appel.

Goodwin, B. C. 1984. Changing from an evolutionary to a generative paradigm in biology. In J. W. Pollard (Ed.), *Evolutionary Theory: Paths into the Future*, 99-120. Chichester and New York: John Wiley and Sons.

Hanson, N. R. 1958. *Patterns of Discovery*. Cambridge: Cambridge University Press.

Huxley, T. H. 1868. On the Animals which are most nearly inter mediate between Birds and Reptiles. M. Foster and E. R. Lankaster, 1898-1902. *The Scientific Memoirs of Thomas Henry*

Huxley, III: 303-313. London: Macmillan.

Huxley, T. H. 1880. The Coming of Age of the Origin of Species. In Foster and Lankaster, 1898-1902. *The Scientific Memoirs of Thomas Henry Huxley*, IV: 395-403. London: Macmillan.

Huxley, T. H. 1894. "Owen's Position in the History of Natural Science", In Rev. R. Owen, *The Life of Richard Owen*, II: 273-332. London: Murray.

Laurent, G. 1987. *Paleontologie et Evolution en France de 1800 a 1860; Une Histoire des Idees de Cuvier et Lamark a Darwin.* Paris: Comite des Travaux Historiques et Scientifiques.

Owen, R. 1846. Report on the Archetype and Homologies of the Vertebrate Skeleton, 169-340. *Report of the British Association for the Advancement of Science (Southampton Meeting).*

Owen, R. 1849. *On the Nature of Limbs.* London: van Voorst.

Owen, R. 1860. *Palaeontology, or a Systematic Summary of Extinct Animals and their Geological Relations*, 63. Edinburgh: Black. Quoted in Desmond, 1982.

Panchen, A. 1992. *Classification, Evolution, and the Nature of Biology.* Cambridge: Cambridge University Press.

Patterson, C. 1982. Morphological characters and homology. In K. A. Joysey and A. E. Friday (Eds.), *Problems in Phylogenetic Reconstruction*, 21-74. London and New York: Academic Press.

Rieppel, O. C. 1988. *Fundamentals of Comparative Biology.* Basel, Boston, Berlin: Birkhauser Verlag.

3

Theoretical Pluralism in Biology, Including Systematics

John Beatty

Department of Ecology, Evolution and Behavior
University of Minnesota
St. Paul, Minnesota 55108

Abstract. In this paper, the notion of "theoretical pluralism" is introduced and discussed in connection with systematics, and biology in general. Theoretical pluralism describes the situation that obtains when no single explanation will suffice for a particular domain of phenomena. The prevalence of theoretical pluralism in biology is evidenced by the preponderance of "relative significance" disputes that biologists wage. In a relative significance dispute, the issue is not *which* theory accounts for a domain of phenomena, but rather the *extent of applicability* of a theory within a domain. Theoretical pluralism and relative significance disputes in systematics and other representative areas of biology are discussed in relation to evolution. In a nutshell, the author argues that the evolutionary history of life would have to have been unrealistically constrained for theoretical unity (instead of theoretical pluralism) to prevail in systematics or any other area of biology.

Copyright © 1994 by Academic Press, Inc.
All rights of reproduction in any form reserved.

WHETHER ONE CELEBRATES IT OR DENIGRATES IT, one can hardly deny the widespread existence of "theoretical pluralism" in biology. What I mean by that (roughly, and for the time being) is that multiple, alternative explanations are *required* for almost every domain of phenomena in biology. This seems on the face of it to be at odds with a long-standing scientific ideal: namely, that there should be a single, unifying explanation for every domain of phenomena.

In this chapter I will explain and illustrate the notion of theoretical pluralism, and discuss how it is possibly at odds with the ideal just mentioned. I will also discuss a very common form of controversy in biology–the so-called "relative significance" dispute–to which theoretical pluralism gives rise. I will then consider why, from an evolutionary point of view, theoretical pluralism is so prevalent in biology in general, and in systematics in particular. In a nutshell: the evolutionary history of life would have to be unrealistically constrained in order for theoretical unity to prevail in any area of biology.

I emphasize from the outset that theoretical pluralism and relative significance disputes are common (inevitable?) *in biology*. Discussion of these issues is usually confined to the more historical and evolutionary sciences: for example, biogeography, systematics, evolutionary biology and ecology. I will place more emphasis on these areas. But in fact, theoretical pluralism and relative significance controversies are prevalent throughout biology; they are not unique to the more historical and evolutionary branches. The same evolutionary reasoning that explains their prevalence in, for example, evolutionary biology itself, also explains their prevalence in, for example, molecular genetics.

Just as molecular geneticists are taking a more evolutionary approach to their subject matter, some systematists are insisting that evolutionary reasoning has played too great a role in their field. The proper subject matter of systematics, they argue, is "pattern," not evolutionary "process" (e.g., Nelson and Platnick, 1981a, 1984; Nelson, 1985; Platnick, 1985; Patterson, 1980, 1982a, 1982b, 1988; Brady, 1985).

Some of these "pattern cladists" insist that the search for natural groups can and should be carried out independently of any evolutionary considerations about how natural groups come into being, and about how patterns of resemblance within and between groups come about. Indeed, they argue that the discovery of natural groups and their relationships is a necessary first step to evolutionary biology; it is the "primary and independent," foundational "prerequisite" for evolutionary thought (Nelson and Platnick, 1981a, p. 283 and throughout).

Pattern cladists are right to warn that circular reasoning can result if the findings of systematics are used to test evolutionary hypotheses that play a role in the determination of those very findings. Thus they are right to argue that a particular grouping cannot be used to test an evolutionary hypothesis if the hypothesis in question has been used to discover or establish the existence of that particular group. But it certainly does not follow that reliance on evolutionary assumptions in systematics *inevitably* leads to circular reasoning. For example, it does not follow that groups identified on the basis of *some* evolutionary assumptions cannot be used to test *any other* evolutionary assumptions. It is simply a non sequitur to argue that "if classifications are to provide an adequate test of theories of the evolutionary process, their construction must be independent of *any theory of process*" (Platnick, 1980, p. 539, my emphasis).

Contrary to the view that systematics is epistemologically prior to evolutionary biology (or that any area of biology is epistemologically prior to any other), I would argue that systematics and evolutionary biology (and every other area of biology) are *reciprocally* illuminating. (See also Hull, 1988a, 1988b; Ridley, 1986; de Queiroz, 1985; de Queiroz and Donoghue, 1990; Sober, 1988; Kluge, 1985; Wiley, 1981; Felsenstein, 1978; and Beatty, 1982, among many others, who argue that while systematics clearly illuminates evolutionary biology, evolutionary reasoning in turn is required to justify the use of various methods, principles, and concepts of systematics.)

There are many respects in which evolutionary reasoning illuminates the practice of systematics. Most importantly for the purposes of this chapter, evolutionary reasoning helps to make sense of why systematics is no more immune from theoretical pluralism and relative significance issues than is any other field of biology–molecular biology or evolutionary biology itself. Evolutionary reasoning thus casts doubt on the attempts of some pattern cladists to establish unitary accounts of several domains–e.g., *the* correct explanation of the relationship between homology and character distributions, *the* correct account of the relationship between ontogeny and character distributions, and *the* appropriate approach to biogeographic patterns. Again, for reasons to be discussed, the evolutionary history of life would have to be unrealistically constrained in order for such theoretical unity to prevail in systematics, or any area of biology.

THEORETICAL PLURALISM AND RELATIVE SIGNIFICANCE DISPUTES

"Pluralism" is a term that already means many (often negative) things to many people. And so I think it is best to begin by making clear what I do *not* mean by that term here. "Pluralism" has a particular meaning in systematics that I especially want to steer clear of for the purposes of this chapter. One sometimes hears that we need more than one species concept, and more than one approach to classification, because species concepts and classifications serve *multiple aims*. Thus, Kitcher (1984) argues that we need "structural" in addition to "historical" species concepts in part to service structural as well as historical inquiries. Similarly, Mayr (1981) argues that classifications should be based on phenetic as well as cladistic methods of analysis in order to portray two different types of information: degree of divergence, and order of splitting. The merits of these arguments are not in question here. The point is just that "theoretical pluralism", as I use the term, does not concern multiple *aims*; rather, it concerns multiple *theories*.

Even "*theoretical* pluralism" has multiple meanings. One that is quite defensible, but different from the sense that I will be discussing, has to do with the idea that there are multiple causes for any particular biological phenomenon. For example, (1) a particular phenotypic trait is the result of the interaction of genotype *and* environment, and (2) no evolutionary change is the result of natural selection alone–in any finite population drift plays *some* role (see Mitchell, 1992, for an interesting discussion of pluralism in this sense).

By "theoretical pluralism," I mean to refer to the way in which biologists explain a *domain* of phenomena, rather than any individual phenomenon. Specifically, theoretical pluralism obtains when the evidence suggests that a multiplicity of theories or mechanisms is needed to explain a domain of phenomena, *different items in the domain requiring explanation in terms of different theories or mechanisms.* There simply is no unitary or unified account of the domain. This is not merely a matter of insufficient evidence for a single theory; rather, it is a matter of the evidence indicating that multiple accounts are required.

Just one example for the time being: let gene regulation be the domain of phenomena. At present we require multiple theories or mechanisms to account for that domain (e.g., Lewin, 1990, pp. 240-299). The well-known lac-operon (negative induction) theory will not account for all cases of gene regulation. We also need positive induction, positive and negative repression, and attenuation models. To say that molecular biologists have found multiple mechanisms of gene regulation is not to say that evidential standards are lax and that "anything goes" in molecular biology. Rather, the evidence strongly suggests that the domain of gene regulation is heterogeneous in the sense that no single existing theory of gene regulation will cover it.

Theoretical pluralism contributes to, and is reflected by, a certain kind of controversy–the so-called "relative significance" dispute. What is at issue in a relative significance dispute is the *extent of applicability* of a theory or mechanism within a domain–roughly, the proportion of items of the domain governed by the theory or mechanism–*not* whether the theory or mechanism in question is *the correct* account of the domain. Thus, for example, molecular biologists argue about the extent of applicability of the lac operon model of gene regulation–*not* whether it is right or wrong, but rather *how commonly* it applies (e.g., Lewin, 1990, pp. 240-299; Yanofsky, 1988).

Examples of theoretical pluralism and relative significance controversies occur at every level of investigation in biology. For instance, immunologists and geneticists argue about the extent of applicability of alternative accounts of the generation of antibody diversity: for example, germ-line versus somatic-cell theories (e.g., Kindt and Capra, 1984). Some geneticists raise questions about the ubiquity of the Mendelian mechanism of inheritance, suggesting that non-Mendelian mechanisms are more prevalent than commonly acknowledged (e.g., Crow, 1979). Physiologists, biophysicists, geneticists, and evolutionary biologists working in the area of gerontology argue about the relative applicability of different mechanisms of aging: that is, somatic mutations in dividing cells versus "wear and tear" of post-mitotic cells (e.g., Comfort, 1979; Maynard Smith, 1966; Rose, 1985; Finch, 1990).

Evolutionary biology, ecology, biogeography and systematics are rife with relative significance controversies. For instance, evolutionary biologists argue about whether selectionist theories have greater applicability to microevolutionary changes than neutralist theories (e.g., Lewontin, 1974; Kimura, 1983; Endler, 1986; Gillespie, 1991). They argue about whether gradualist, adaptationist theories of macroevolution have greater applicability than the punctuated equilibrium theory (e.g., Gould, 1980; Lande, 1980). They argue about the extent of applicability of the various mechanisms of the evolution of sex, from the "red-queen" hypothesis to the "tangled-bank" hypothesis to the "genetic-load" model to the "DNA-repair" model (e.g., Kondrashov, 1988).

Evolutionary biologists and systematists argue about the extent of applicability of each of the multitude of theories of speciation, from each of the various forms of sympatric speciation, to parapatric speciation, to each of the various forms of allopatric speciation (e.g., Bush, 1975; White, 1978; Otte and Endler, 1989). A related area of evolutionary biology and systematics where theoretical pluralism *might* be considered prevalent has to do with species concepts (the emphasis indicates that I am not certain whether the pluralism of species concepts is appropriately construed as "theoretical" pluralism). One might argue that debates over the variety of competing species concepts are debates about definitions, not theories. And to a certain extent the arguments are certainly about convention as well as empirical truth. But the debates are also theoretical: about what sorts of entities, above the level of organisms, evolution has produced, and about what sorts of entities evolve. The multiplicity of species concepts also reflects in part the multiplicity of theories of speciation (on the pluralism of species concepts and the theoretical, empirical, and conventional issues involved, see Mishler and Donoghue, 1982; Kitcher, 1984; Mishler and Brandon, 1987; de Queiroz and Donoghue, 1988; and Ereshefsky, 1992).

Evolutionary biologists and systematists argue about the relative significance of different accounts of the relationship between ontogeny and phylogeny (e.g., the relative applicability of "von Baer's law"; see, e.g., Gould, 1977). They argue about whether vicariance accounts of biogeographic patterns have greater applicability than dispersalist/center-of-origin accounts (e.g., Nelson and Platnick, 1981a, 1981b).

Ecologists debate the extent of applicability of alternative theories of community structure, from competition theory, to predation and abiotic factor theories, to random colonization models (e.g., Schoener, 1982; Connell, 1983; Schoener, 1983; Sih et al., 1985). Again, these are all disputes about the extent of applicability of alternative theories or mechanisms within a particular domain, *not* whether this or that account is the universally true one within that range.

As I will explain later, the extent of theoretical pluralism and relative significance issues in biology is not a mystery. From an evolutionary point of view, it would be most remarkable if theoretical unity prevailed instead.

THEORETICAL PLURALISM AND RELATIVE SIGNIFICANCE IN SYSTEMATICS

Theoretical pluralism may not seem relevant to systematics, because of the belief on the part of many that systematics is, or ought to be, theory neutral or theory minimal. Were systematics actually theory neutral or theory minimal, then it would be immune, or relatively immune, from theoretical pluralism and relative significance issues. It is a common enough conception that systematics is atheoretical–common enough that Mayr often attacked the notion in defense of systematics: e.g., "Let us remember that taxonomy is not a kind of stamp-collecting but a branch of biology," not purely "descriptive," but every bit as "explanatory" as any other area of biology (e.g., Mayr, 1968, p. 599).

Nevertheless, some systematists promote attempts to make the field theory neutral or at least theory minimal, rendering it not so much a branch of biology as the epistemological trunk. This was a major motivation behind phenetics, and it is now a major motivation behind "pattern cladism" (Hull, 1988a). Pattern cladists emphasize a distinction between theory neutral or theory minimal *patterns*, and theoretical *process*-accounts of those patterns. They are especially concerned to free systematics of as many *evolutionary* process assumptions as possible; some aim to rid systematics of all evolutionary assumptions. For example, according to Patterson,

> As the theory of cladistics has developed, it has been realized that more and more of the evolutionary framework is inessential, and may be dropped . . . Platnick refers to the new theory as "transformed cladistics" and the transformation is away from dependence on evolutionary theory. Indeed, Gareth Nelson, who is chiefly responsible for the transformation, put it like this in a letter to me this summer: "In a way, I think we are rediscovering preevolutionary systematics; or if not rediscovering it, fleshing it out." (Patterson, 1980, p. 239)

The discovery of pattern, pattern cladists argue, is the central–indeed, epistemologically prior–task of systematics. Reasoning about evolutionary processes does not illuminate, but rather is illuminated by, pattern analysis. According to Nelson and Platnick,

> Patterns...are interesting for [various] reasons. One is the fashion these days to discuss processes of evolution. Some biologists expound on processes as if all worthwhile general knowledge is contained therein. Now what, one might ask, are processes of evolution? Do they not all presuppose the existence of a nonrandom pattern such as the one we have considered? No patterns–in general, no processes. No patterns, nothing to explain by invoking one or another concept of process. In short, a process is that which is the cause of a pattern. No more, no less. Pattern analysis is, in its own right, both primary and independent of theories of process, and is a necessary prerequisite to any analysis of process. (Nelson and Platnick, 1981a, p. 35).

Pattern cladists further argue that pattern analysis is best carried out by using what are supposed to be evolutionarily neutral "methods," such as "parsimony" and "ontogenetic" criteria. As Nelson and Platnick say of parsimony, "the use of a parsimony criterion does not presuppose anything in particular about the nature of evolution" (1981a, p. 222). And as Patterson says of the ontogenetic criterion,

> ...there is one apparently foolproof method: it depends not on "evolution" as we understand it, but on the usage of that word among pre-Darwinian biologists. To them, "evolution" meant the unfolding of pattern, increase in complexity, or differentiation seen in the development of individual organisms—the process of embryology rather than descent with modification. (Patterson, 1982b, p. 304)

But as many authors have argued, these criteria can only be justified by evolutionary reasoning. That is, one can adopt these methods for no reason, for bad reasons, for grossly incomplete reasons, or for reasons that are in part evolutionary. Moreover, the evolutionary grounds that justify these methods do not justify them universally; but rather only in some circumstances. The methods should be judged, then, in terms of the *extent* of their applicability, not their correctness per se. Insofar as these methods play a role in explaining patterns of similarity and dissimilarity among organisms (character distributions), systematics bears all the marks of theoretical pluralism and relative significance issues that are characteristic of every other branch of biology.

The role that methods such as parsimony and ontogenetic criteria play in understanding character distributions derives from their role in choosing between different phylogenetic trees, cladograms and groupings. Take for instance the following version of parsimony: the phylogenetic tree, cladogram or grouping that best accounts for a particular character distribution is the one that requires the fewest non-homologous similarities, or the fewest evolutionary character changes (Sober, 1988, p. 31). It is interesting that Nelson and Platnick (1981a, pp. 201, 213) themselves articulate parsimony criteria as injunctions to choose the trees or groupings that "imply" the fewest evolutionary changes.

The parsimony criterion, thus construed, plays an important role in explanations of character distributions. An ideally complete account of a character distribution would include the most parsimonious grouping hypothesis consistent with the character distribution, together with the evolutionary assumptions used to justify the parsimony criterion–in other words, the reasons for thinking that the most parsimonious grouping hypothesis is the best way to make sense of the data. Preferably, these reasons for using parsimony would be made explicit, and ideally they would be supported by independent evidence. Of course, there is considerable discussion and debate at present about the precise evolutionary conditions under which parsimony is justified. But there is widespread agreement that *some* evolutionary conditions are assumed in justifying its use (e.g., Sober, 1988; Donoghue, 1990; and references in the next paragraph; see especially the pioneering work of Felsenstein, 1978).

There are a number of different parsimony criteria–e.g., Fitch, Wagner, Dollo, and Camin-Sokal parsimony criteria. There are also a number of different alternatives to parsimony–alternative criteria for choosing which phylogenetic hypothesis best accounts for a character distribution–e.g., compatibility, distance, and maximum likelihood methods. These alternative criteria, through their alternative evolutionary justifications, correspond to alternative theoretical explanation-types of the domain of character distributions (although, to be precise, the maximum likelihood approach is not associated with any particular set of evolutionary assumptions, but rather requires for its implementation the formulation of a precise evolutionary account of the data set). See the review of alternative methods, and their alternative evolutionary assumptions, in Felsenstein (1982, 1988) and Swofford and Olsen (1990). As Maddison and Maddison summarize, "Each of these methods can succeed or fail, depending on the nature of the evolutionary process" (1992, p. 51).

There is therefore no issue as to which of these theoretical explanation-types is *the* way to account for the domain. At issue, instead, is the relative significance of the various grouping criteria, and hence of the various kinds of accounts that they enter into. The most that one hopes for with regard to parsimony criteria, for instance, is that the character distributions being investigated "reside in that sizeable portion of the universe where parsimony is justified" (Donoghue, 1990, p. 123). Enterprising efforts are currently under way to estimate the ranges of applicability of the various grouping criteria within the domain of four-taxon character distributions (based ultimately on Felsenstein, 1978).

It is worth considering one other frequently discussed category of phylogeny-choice or grouping criteria–namely, "ontogenetic" criteria. These are not necessarily alternatives to parsimony criteria; rather they are usually construed as complementary, and used in conjunction with parsimony criteria. Generally speaking, according to these criteria we should choose the phylogenetic tree, cladogram, or grouping hypothesis that best accounts not only for the character distributions in question (where the character distributions are traditionally conceived), but also similarities and differences between ontogenetic stages and sequences of the organisms in question.

One version of the ontogenetic criterion is associated with "von Baer's law," according to which, "The general features of a large group of animals appear earlier in the embryo than the special features" (quoted and translated in Gould, 1977, p. 56). The ontogenetic criterion associated with this law equates the "general" characters with the evolutionarily more primitive, and the less general or "special" with the evolutionarily more derived characters. The criterion thus licenses us to infer from the ontogenetic sequence x → y that x is the more general character and thus, on grounds of parsimony, that x is the primitive character, and y derived. We should thus reject a grouping hypothesis that would require y to appear before x in ancestors of the taxon in question. To proceed otherwise would require an additional evolutionary stage (y → x → y), and hence would require unparsimoniously many independent origins of the derived character.

As numerous authors have reasoned, the use of this criterion depends on the evolutionary assumption of "terminal addition," i.e., that new, derived characters of descendant taxa appear at the end of the ontogenies of ancestral taxa, rather than being inserted into earlier stages of the ancestral ontogeny, and that stages of development of ancestral taxa are not lost in descendant taxa (the phenomena ruled out are called "non-terminal insertion," "terminal deletion," and "non-terminal deletion."

A complete explanation of a character distribution in terms of the grouping hypothesis that best satisfies the ontogenetic criterion would include the evolutionary assumptions used to justify the criterion (preferably explicitly, and ideally along with independent evidence that the assumptions hold in this case).

The ontogenetic criterion associated with von Baer's law is considered to be somewhat applicable, but by no means universally applicable within the domain of character (and ontogeny) distributions. von Baer's "law" is not a law–it has numerous alleged exceptions (de Beer, 1940; Gould, 1977; Rieppel, 1979; Stevens, 1980; Fink, 1982; Kluge, 1985). Thus the issue with regard to this criterion type is its relative significance. Ridley cites authors who believe that terminal addition is common, and other authors who believe it is infrequent: "We can only conclude that terminal addition is at least *sufficiently common* for von Baer's law to possess *some* use for the cladist (Ridley, 1986, p. 67).

Pattern cladists have defended a different version of the ontogenetic criterion, which they defend as an exceptionless law, independent of evolutionary assumptions. This is Nelson's "biogenetic law:" "Given an ontogenetic character transformation, from a character observed to be more general to a character observed to be less general, the more general character is primitive and the less general advanced" (Nelson, 1978, p. 327).

The careful phrasing of the antecedent clause, "Given an ontogenetic character transformation from a character observed to be more general to a character observed to be less general," does protect the generalization from cases that would falsify von Baer's law. Nelson's law does not apply in cases where more general characters arise in ontogeny from less general.

However, that does not protect the "law" from all evolutionarily possible circumstances. As de Queiroz (1985) points out, the truth of the generalization rests on whether ancestral states are retained in descendant ontogenies. So, for instance, deletion of an ancestral character from the ontogeny of descendant taxa is ruled out if the generalization is really true. Moreover, de Queiroz's conception of the "generality" of a character is charitable to Nelson. Using Nelson's own criterion of generality, the biogenetic "law" could be falsified not only by terminal or non-terminal deletion, but also by non-terminal insertion. The truth of Nelson's "law" rests on terminal addition in evolution every bit as much as the truth of von Baer's.

But that is only to say that Nelson's "law" has potential falsifiers. Nelson and Platnick suggested that apparent cases of terminal deletion are "not a falsifier at all, but only a reflection of lack of information. One may doubt, for example, that any characters are truly lost rather than transformed. Apparent loss may be an indication that the characters and transformations are merely poorly understood and, consequently wrongly defined" (Nelson and Platnick, 1981a, p. 353). Nonetheless, numerous theoretical analyses and empirical studies suggest otherwise (Lundberg, 1973; de Queiroz, 1985; Kluge, 1985; Mabee, 1989; see Weston, 1988, for a dissenting view). As Mabee summarizes, whether the ontogenetic criterion (Nelson's version) is illuminating or misleading "is thus dependent upon the phylogeny from which the polarity [judgment as to which character is primitive and which derived] is inferred" (Mabee, 1989, p. 410). In other words, the issue is not whether Nelson's "law" is *the* account of the domain specified in his antecedent clause; rather, the issue concerns the *relative significance* of terminal addition within that domain.

TEMPERING THEORETICAL PLURALISM: THE NEWTONIAN TRADITION

The fact that theoretical pluralism and relative significance disputes are so common in biology contrasts strikingly, I believe, with a traditional ideal in the physical sciences, namely, the aim to explain a domain of phenomena in terms of as few as possible different mechanisms, and best of all one single mechanism. This ideal was expressed particularly well by Newton, and so I will call it the

"Newtonian tradition." Newton elaborated it most succinctly in the first two of his three "rules of reasoning in philosophy" (Newton, 1686). According to the first rule, "We are to admit no more causes of natural things than such as are both true and sufficient to explain their appearances." As Newton clarified the rule, "To this purpose the philosophers say that Nature does nothing in vain, and more is in vain when less will serve; for Nature is pleased with simplicity and affects not the pomp of superfluous causes." Newton's second rule states my point more clearly: "Therefore [i.e., it follows from the first rule that] to the same natural effects we must, as far as possible, assign the same causes." As he proceeds to illustrate the rule: "As to respiration in a man and in a beast, the descent of stones in Europe and America, the light of our culinary fire and of the sun, the reflection of light in the earth and in the planets."

Judging by their acceptance of theoretical pluralism, and by their waging of relative significance disputes, biologists seem not overly impressed by this rule of reasoning. Indeed, by their *promotion* of theoretical pluralism they seem to *repudiate* the Newtonian ideal.

For example, in their reviews of the modes of speciation, Bush (1975) and White (1978) staunchly defend a pluralistic approach against assumptions or attempts to show that there is a single correct account of the domain. As White insists,

> However much evolutionists of the future may synthesize in the field of speciation, we can be confident that the diversity of living organisms is such that their evolutionary mechanisms cannot be forced into the straightjacket of any narrow, universal dogma (White, 1978, p. 349).

This pluralism is also characteristic of the recent anthology and state-of-the-art summary, *Speciation and its Consequences*, edited by Otte and Endler (1989).

McIntosh (1987) recently summarized the trend toward theoretical pluralism in ecology, away from the ideals of the sixties and early seventies when ecologists like MacArthur envisioned that all of ecology would ultimately be "embodied in a small number of simple laws." The most recent anthologies, for example the anthology on community ecology edited by Diamond and Case (1986), proclaim pluralism in the preface and throughout. The editors explicitly distance themselves from the ideals of Newtonian mechanics:

> Until recently, philosophy of science focused on relatively homogeneous fields such as classical mechanics. As a result, many scientists have been trained to regard pluralistic approaches as soft, unrigorous, unscientific, and indicative of a retarded field. Even scientists who work in pluralistic fields tend to view how science "should" be pursued in ways that are mismatched to their field's special needs. (Diamond and Case, 1986)

"Does this apply always, sometimes, or never?"

©1977 by Sidney Harris — "What's So Funny About Science?", William Kaufmann Inc.

The Newtonian tradition may prevail more in the physical sciences (at least in the nonhistorical–e.g., nongeological, noncosmological, physical sciences). The difference between that tradition, and the tradition of relative significance controversies that prevails in biology, is well illustrated by the Sidney Harris cartoon (previous page) of two physicists (they're not mathematicians–mathematicians don't wear white coats).

The assumption behind the cartoon–what makes it funny–is that physicists are not supposed to argue about such matters. But what makes us think these are physicists? Well, if they were not, it would not be funny. Imagine that they are evolutionists/systematists arguing about modes of speciation:

"Does this apply always, sometimes, or never?"

©1977 by Sidney Harris – "What's So Funny About Science?", William Kaufmann Inc.

or systematists arguing about the circumstances under which parsimony is misleading:

"Does this apply always, sometimes, or never?"

©1977 by Sidney Harris – "What's So Funny About Science?", William Kaufmann Inc.

Now this is not a joke. It is rather the fact of the matter. To some it is the sad fact of the matter. Which leads me to temper my remarks about theoretical pluralism in biology.

It is important not to exaggerate the differences between the biological and physical sciences. The Newtonian tradition has considerable appeal in biology as well, and not only in the more reductionistic branches of biology, like molecular biology. One also finds it in systematics, ecology and evolutionary biology.

For example, it would seem to be the Newtonian ideal that explains why some pattern cladists like Nelson and Platnick regard "Darwinism" as simply false and dispensable, and why they neglect or decline to evaluate instead the *relative significance* of the Darwinian approach:

> With respect to the overall status of Darwinism, which has been much debated in recent times, there is little to say. The intrusion of Darwinism into systematics was at worst a misadventure, but it was not without value. Science learns through mistakes, and science may count as progress the falsification of a theory, even if in retrospect the theory seems to have been wrongheaded from the beginning. (Nelson and Platnick, 1984, p. 145)

One wonders, though, whether we would be left with *any* theories in biology if we evaluated them all as either universally true within a domain, or false and dispensable.

The Newtonian tradition is more often manifested in biology as an *ideal.* That is, theoretical pluralism, and the necessity of assessing the relative significance of theories, is often acknowledged *for the time being*, but only as a temporary malady. For example, Carson chides his pluralistic peers in the area of speciation for giving up too easily in this regard:

> Despite much modern work in plant and animal population biology, there has been a regrettable lack of unification of theory relating to the modes or processes involved in the origin of new species. I find two reasons for this. First, there is a tendency not to be reductionistic, that is, to accept many disparate theoretical notions about the way in which species may arise (e.g., White, 1978). In the face of this, long and complex classifications of various conceivable modes of speciation (e.g., allopatric, sympatric, parapatric, stasipatric, etc.) have been constructed, discouraging those who seek unifying principles. (Carson, 1985, p. 380)

Many biogeographers, systematists, ecologists and evolutionary biologists reveal the limits of their tolerance for theoretical pluralism by conducting their relative significance arguments in the manner described by Gould and Lewontin:

> In natural history, all possible things happen sometimes; you generally do not support your favoured phenomenon by declaring rivals impossible in theory. Rather, you acknowledge the rival, but circumscribe its domain of action so narrowly that it cannot have any importance in the affairs of nature. Then, you often congratulate yourself for being such an ecumenical chap. (Gould and Lewontin, 1979, p. 585)

To a certain extent, as Gould and Lewontin suggest, biogeographers, systematists, ecologists, and evolutionary biologists often acknowledge the need for theoretical pluralism, but try to keep it under control by minimizing the significance of all but a couple, or even one, possible account of a domain of phenomena.

But the fact that tactics like these are employed to contain theoretical pluralism indicates that theoretical pluralism is indeed widespread, however much some biologists with Newtonian inclinations may regret it. As Gilpin acknowledges in objecting to a fellow ecologist's pluralistic approach:

> I must confess that I am saddened by [his book's] honest realism, its unabashed pluralism. Something of a romantic, I long for the heady days of an earlier decade when the [alternatives considered by him] vied one against the other to be the organizing principles of our science. (Gilpin, 1986, pp. 200-201)

THEORETICAL PLURALISM FROM AN EVOLUTIONARY PERSPECTIVE

Is there any reason to believe that theoretical pluralism is just a temporary malady? Alternatively, is there any reason–beyond faith or aesthetics–to expect that we will someday find unitary accounts to replace the pluralistic accounts that prevail at present?

The answer to that question should be evolutionarily informed (not *merely* evolutionarily informed, but at least *in part* evolutionarily informed). The biological world is, after all, an outcome of evolution; true biological generalizations are descriptions of evolutionary outcomes.

Thus, for example, if Mendel's theory alone sufficed to account for inheritance in sexual organisms–that is, if Mendel's "law" of segregation correctly described inheritance at all loci of all sexual organisms–that would be a fact that we ought (in principle) to be able to explain evolutionarily.

Mendelian inheritance is quite prevalent, and there are evolutionary biologists who are trying to explain that fact (e.g., Bell, 1982; Uyenoyama, 1987). But Mendelian inheritance is not characteristic of every locus of every species of sexual organism–not even close. Non-Mendelian mechanisms are also prevalent; and the prevalence of each non-Mendelian mechanism also requires an evolutionary account. (Actually as Bell has acknowledged, many evolutionary biologists find it easier to explain the prevalence of non-Mendelian than Mendelian mechanisms of inheritance–Bell, 1982, p. 439.)

To discover a unitary account of a domain of biological phenomena (e.g., inheritance, or development, or foraging behavior), we would have to sample the diversity of taxa covered by each domain (e.g., sexual organisms, eukaryotes, or herbivores) very thoroughly, and find a single account for all the items in our sample. And because even the most comprehensive sample would still be very incomplete, we would also want some evolutionary justification for our extrapolation. That is, we would like to have some reason for believing that in this case–unlike so many others–evolution had actually resulted in only one outcome: one mechanism of inheritance in sexual organisms, one mechanism of development in eukaryotes, one foraging strategy in herbivores.

But why would we expect evolutionary outcomes to be so constrained? *Extreme* phylogenetic conservativism? *Extremely* strong and remarkably similar selection pressures resulting in *extreme* convergence? In other words, to suggest a single, common mechanism underlying a domain of biological phenomena, we must assume that the mechanism evolved in a common ancestor of all the taxa covered by the domain, and has been maintained ever since, or that the very same mechanism has arisen independently in all the taxa covered by the domain (or some combination of the two explanations). But those are hardly *a priori* obvious assumptions.

Why would we expect evolution to result in a single mechanism of gene regulation (for a discussion of the evolution of regulatory genes and patterns, see Dickinson, 1991)? Why would we expect to find a single mechanism of aging (for a discussion of the evolution of different mechanisms and patterns of aging, see Finch, 1990)? If we thought that sex had arisen only once, then it would be reasonable to aim for a single account of the *origin* of sex. But even in that case, why would we expect to find a single account of the evolutionary *maintenance* of sex in all the independently evolving sexual lineages (see the references listed above)?

We certainly have no evolutionary reason for expecting a unitary account of speciation. Just the opposite. Evolution has resulted in a diverse array of genetic systems and mating systems, which are differentially susceptible to different mechanisms of speciation (e.g., Bush, 1975; White, 1978).

Similarly, we have no good evolutionary reason for expecting a unitary account of the relationship between ontogeny, character distribution, and phylogeny. As discussed above, both von Baer's law and Nelson's hold only when the evolution of ontogenies proceeds in a particular manner. There is no reason to expect either principle to hold universally if evolution does not always proceed in this way. To adhere to either "law" as *the* account of the relationship between ontogeny and phylogeny (or *the* account of the relationship between ontogeny and patterns of character distribution) would be to place great faith in the idea that evolution could not result in any other ontogenetic outcome.

CONCLUDING QUESTIONS

We need evolutionary reasons to be Newtonians in biology. But general evolutionary grounds for Newtonian thinking in biology are not easy to find and defend. The point of the present discussion is not so much to offer evolutionary reasons for expecting disunity, as it is to raise doubts about whether we have reasons to expect unity to prevail, and to point out that evolutionary reasoning is *relevant* to the issue of whether unitary or pluralistic accounts of the domains of biology will eventually be found.

But I can imagine an argument for adhering to the Newtonian ideal, without invoking evolutionary reasoning in the process. That is, one might argue, whether theoretical pluralism reflects the nature of the biological world, or the state of our ignorance (our ignorance of the unitary or unifying theory for each domain of biological phenomena), we cannot at present know. Nonetheless, we should aim for unitary or unifying theories. Then, if the biological world is really inescapably heterogeneous, we will ultimately be forced to deal with theoretical pluralism. But if we begin by assuming theoretical pluralism, then we may never find the unitary or unifying theories that might actually be true. We might rest happy with multiple accounts when a unitary account is possible and could be discovered with just a little more effort. I am not just imagining this argument; it is a lot like Hull's argument in favor of a single species concept. According to Hull, defenders of a unitary versus a pluralistic species concept

> ...are carrying on in the best scientific tradition of opting for one perspective and pushing it for all its worth. Perhaps species as genealogical actors in an ecological play may prove ultimately inadequate. Science does march on. If so, then this monism will have proved to have been only temporary, but the only way to find out how adequate a particular conception happens to be is to give it a run for its money. Remaining content with a variety of slightly or radically different species concepts might be admirably open-minded and liberal, but it would be destructive of science.... (Hull, 1987, p. 178)

To elaborate a bit on the argument, *if* nature is inescapably heterogeneous, the Newtonian would not forever overlook that fact, but would be faced with it over and over again. The Newtonian would ultimately be forced to acknowledge theoretical pluralism in that case. According to this argument, theoretical pluralism is possibly misleading, while the Newtonian tradition is, at worst, inefficient.

This is a difficult argument to counter. The best I can do is to offer an alternative argument (rather, sketch of an argument), which rests on the following premise. Just as the various areas of biology stand to one another in the relationship of reciprocal illumination, so do biological methodology and our best biological theories. That is, biological methodology, including injunctions to seek unified accounts of each and every domain, should be biologically (in this case evolutionarily) informed. This is the argument that structures Sober's analysis of the evolutionary grounds underlying the parsimony criterion in systematics. As Sober put it, "The idea of a presuppositionless 'scientific method' implies that methodology is static and insensitive to what we learn about the world. But with theory and method linked by a subtle nexus of interdependence, progress on theories can be expected to improve our methods of inquiry" (Sober, 1988, p. 239).

Similarly, why should we adhere to a methodology that dictates the search for unitary accounts of each domain of biological phenomena–e.g., a unitary account of inheritance, or a unitary account of gene regulation, or a unitary account of speciation, or a unitary account of the relationship between ontogeny and character distributions–unless we have reason to believe that the outcomes of evolution are *highly* constrained? If we doubt that the outcomes of evolution are so highly constrained, then perhaps we should be *on the lookout* for multiple accounts of each domain. Only a naive Newtonian would rest satisfied with a unitary account, when, with a little more effort, a multiplicity of accounts might be found!

Is the concern to discover *the* account of every domain in biology a matter of ideals, regardless of the nature of the biological world? Is the concern to discover *the* account of every domain in biology a *biologically* informed position?

Acknowledgments

I am very grateful for criticisms and suggestions received from Michael Donoghue, Lance Grande, Olivier Rieppel, and three anonymous reviewers.

References

Beatty, J. 1982. Classes and cladists. *Systematic Zoology*, 31: 25-34.

Bell, G. 1982. *The Masterpiece of Nature: The Evolution and Genetics of Sexuality*. Berkeley: University of California Press.

Brady, R. 1985. On the independence of systematics. *Cladistics*, 1: 113-126.

Bush, G. 1975. Modes of animal speciation. *Annual Review of Ecology and Systematics*, 6: 339-364.

Carson, H. L. 1985. Unification of speciation theory in plants and animals. *Systematic Botany*, 10: 380-390.

Comfort, A. 1979. *The Biology of Senescence*. New York: Elsevier.

Connell, J. H. 1983. On the prevalence and relative importance of interspecific competition: evidence from field experiments. *American Naturalist*, 122: 661-696.

Crow, J. 1979. Genes that violate Mendel's rules. *Scientific American*, 240(2): 134-146.

Darwin, C. 1859. *On the Origin of Species*. London: Murray.

de Beer, G. 1940. Embryology and taxonomy. In J. Huxley (Ed.), *The New Systematics*, 365-393. Oxford: Oxford University Press.

de Queiroz, K. 1985. The ontogenetic method for detecting character polarity and its relation to phylogenetic systematics. *Systematic Zoology*, 34: 280-299.

de Queiroz, K., and M. J. Donoghue. 1988. Phylogenetic systematics and the species problem. *Cladistics*, 4: 317-338.

de Queiroz, K., and M. J. Donoghue. 1990. Phylogenetic systematics or Nelson's version of cladistics? *Cladistics*, 6: 61-75.

Diamond, J., and T. Case (Eds.) 1986. *Community Ecology*. New York: Harper and Row.

Dickinson, W. J. 1991. The evolution of regulatory genes and patterns in *Drosophila*. *Evolutionary Biology*, 25: 127-173.

Donoghue, M. J. 1990. Why parsimony? *Evolution*, 44: 1121-1123.

Endler, J. A. 1986. *Natural Selection in the Wild*. Princeton: Princeton University Press.

Ereshefsky, M. 1992. Eliminative pluralism. *Philosophy of Science*.

Felsenstein, J. 1978. Cases in which parsimony or compatibility methods can be positively misleading. *Systematic Zoology*, 27: 401-410.

Felsenstein, J. 1982. Numerical methods for inferring evolutionary trees. *Quarterly Review of Biology*, 57: 379-404.

Felsenstein, J. 1988. Phylogenies from molecular sequences: inference and reliability. *Annual Review of Genetics*, 22: 521-565.

Finch, C. E. 1990. *Longevity, Senescence, and the Genome.* Chicago: University of Chicago Press.

Fink, W. L. 1982. The conceptual relationship between ontogeny and phylogeny. *Paleobiology*, 8: 254-264.

Gillespie, J. H. 1991. *The Causes of Molecular Evolution.* Oxford: Oxford University Press.

Gilpin, M. E. 1986. Review of R. J. Taylor, *Predation*. *American Scientist*, 74(2): 200-201.

Gould, J. L., and C. G. Gould. 1989. *Sexual Selection*. New York: Scientific American Library.

Gould, S. J. 1977. *Ontogeny and Phylogeny.* Cambridge: Harvard University Press.

Gould, S. J. 1980. Is a new and general theory of evolution emerging? *Paleobiology*, 6: 119-130.

Gould, S. J., and R. C. Lewontin. 1979. The spandrels of San Marco and the Panglossian paradigm: a critique of the Adaptationist Programme. *Proceedings of the Royal Society of London*, B205: 581-598.

Harris, S. 1977. *What's So Funny About Science?* Los Altos, CA: Kaufmann.

Hull, D. L. 1987. Genealogical actors in ecological roles. *Biology and Philosophy*, 2: 168-184.

Hull, D. L. 1988a. *Science as a Process: An Evolutionary Account of the Social and Conceptual Development of Science*. Chicago: University of Chicago Press.

Hull, D. L. 1988b. Progress in ideas of progress. In M. Nitecki (Ed.),

Evolutionary Progress. Chicago: University of Chicago Press.
Kimura, M. 1983. *The Neutral Theory of Molecular Evolution*. Cambridge: Cambridge University Press.
Kindt, T. J., and J. C. Capra. 1984. *The Antibody Enigma*. New York: Plenum.
Kitcher, P. 1984. Species. *Philosophy of Science*, 51: 308-333.
Kluge, A. G. 1985. Ontogeny and phylogenetic systematics. *Cladistics*, 1: 13-27.
Kluge, A. G., and R. Strauss. 1985. Ontogeny and systematics. *Annual Review of Ecology and Systematics*, 16: 247-268.
Kondrashov, A. S. 1988. Deleterious mutations and the evolution of sexual reproduction. *Nature (London)*, 336: 435-440.
Lande, R. 1980. Microevolution in relation to macroevolution. *Paleobiology*, 6: 235-238.
Lewin, B. 1990. *Genes IV*. Oxford: Oxford University Press.
Lewontin, R. C. 1974. *The Genetic Basis of Evolutionary Change*. New York: Columbia University Press.
Lundberg, J. G. 1973. More on primitiveness, higher level phylogenies and ontogenetic transformations. *Systematic Zoology*, 22: 327-329.
McIntosh, R. P. 1987. Pluralism in ecology. *Annual Review of Ecology and Systematics*, 18: 321-341.
Mabee, P. M. 1989. An empirical rejection of the ontogenetic polarity criterion. *Cladistics*, 5: 409-416.
Maddison, W. P., and D. R. Maddison. 1992. *MacClade. Analysis of Phylogeny and Character Evolution. Version 3*. Sunderland, MA: Sinauer Associates, Inc. 398 pp.
Maddison, W. P., M. J. Donoghue and D. R. Maddison. 1984. Outgroup analysis and parsimony. *Systematic Zoology*, 33: 83-103.
Maynard Smith, J. 1966. Theories of aging. In P. L. Krohn (Ed.), *Topics in the Biology of Aging*. New York: Interscience.
Mayr, E. 1968. The role of systematics in biology. *Science*, 159: 595-599.
Mayr, E. 1981. Biological classification: toward a synthesis of opposing methodologies. *Science*, 214: 510-516.
Michod, R. E., and B. R. Levin (Eds.). 1988. *The Evolution of Sex*.

Sunderland, MA: Sinauer Associates Inc.

Mishler, B., and R. Brandon. 1987. Individuality, pluralism, and the phylogenetic species concept. *Biology and Philosophy*, 2: 145-166.

Mishler, B., and M. Donoghue. 1982. Species concepts: A case for pluralism. *Systematic Zoology*, 31: 491-503.

Mitchell, S. 1992. On pluralism and competition in evolutionary explanations. *American Zoologist*, 32: 135-144.

Nelson, G. 1978. Ontogeny, phylogeny, paleontology and the biogenetic law. *Systematic Zoology*, 27: 324-345.

Nelson, G. 1985. Outgroups and ontogeny. *Cladistics*, 1: 29-45.

Nelson, G., and N. Platnick. 1981a. *Systematics and Biogeography: Cladistics and Vicariance*. New York: Columbia University Press.

Nelson, G., and N. Platnick (Eds.). 1981b. *Vicariance Biogeography: A Critique*. New York: Columbia University Press. 593 pp.

Nelson, G., and N. Platnick. 1984. Systematics and evolution. In M.-W. Ho and P. T. Saunders (Eds.), *Beyond Neo-Darwinism: An Introduction to the New Evolutionary Paradigm*. London: Academic Press. 376 pp.

Newton, I. 1686. *Philosophia Naturalis Principia Mathematica*, Book III.

Otte, D., and J. A. Endler (Eds.). 1989. *Speciation and its Consequences*. Sunderland, MA: Sinauer Associates Inc. 679 pp.

Patterson, C. 1980. Cladistics. *Biologist*, 27: 234-240.

Patterson, C. 1982a. Morphological characters and homology. In K. A. Joysey and A. E. Friday (Eds.), *Problems of Phylogenetic Reconstruction*. London: Academic Press.

Patterson, C. 1982b. Cladistics and classification. *New Scientist*, 94: 303-306.

Patterson, C. 1988. The impact of evolutionary theories on systematics. In D. L. Hawksworth (Ed.), *Prospects in Systematics*. Oxford: Clarendon Press.

Platnick, N. I. 1980. Philosophy and the transformation of cladistics. *Systematic Zoology*, 28: 537-546.

Platnick, N. I. 1985. Philosophy and the transformation of cladistics revisited. *Cladistics*, 1: 87-94.

Ridley, M. 1986. *Evolution and Classification*. London: Longman.

Rieppel, O. 1979. Ontogeny and the recognition of primitive character states. *Zeitschrift für Zoologische Systematik und Evolutionsforschung*, 17: 57-61.

Rose, M. R. 1985. The evolution of senescence. In P. J. Greenwood et al. (Eds.), *Evolution: Essays in Honour of John Maynard Smith*. Cambridge: Cambridge University Press.

Schoener, T. 1982. The controversy over interspecific competition. *American Scientist*, 70: 586-595.

Schoener, T. 1983. Field experiments on interspecific competition. *American Naturalist*, 122: 240-285.

Sih, A., P. Crowley, M. McPeek, J. Petranka and K. Strohmeier. 1985. Predation, competition, and prey communities: a review of field experiments. *Annual Review of Ecology and Systematics*, 16: 269-311.

Sober, E. 1988. *Reconstructing the Past: Parsimony, Evolution, and Inference*. Cambridge, MA: MIT Press.

Stevens, P. F. 1980. Evolutionary polarity of character states. *Annual Review of Ecology and Systematics*, 11: 333-358.

Swofford, D. L., and G. J. Olsen. 1990. Phylogeny reconstruction. In D. M. Hillis and C. Moritz (Eds.), *Molecular Systematics*, 411-501. Sunderland, MA: Sinauer Associates Inc.

Uyenoyama, M. K. 1987. Genetic transmission and the evolution of reproduction: the significance of parent-offspring relatedness to the "cost of meiosis". In P. B. Moens (Ed.), *Meiosis*. Orlando, FL: Academic Press.

Weston, P. H. 1988. Indirect and direct methods in systematics. In C. J. Humphries (Ed.), *Ontogeny and Systematics*. New York: Columbia University Press.

White, J. D. 1978. *Modes of Speciation*. San Francisco: W. H. Freeman.

Wiley, E. O. 1981. *Phylogenetics: The Theory and Practice of Phylogenetic Systematics*. New York: John Wiley and Sons. 439 pp.

Yanofsky, C. 1988. Transcription attenuation. *Journal of Biological Chemistry*, 263: 609-612.

4

Repeating Patterns in Nature, Predictability, and "Impact" in Science

Lance Grande

Department of Geology
Field Museum of Natural History
Chicago, Illinois 60605

Abstract. Some of the most broadly significant findings relevant to evolutionary biology (and to other sciences) are repeating patterns in nature. In phylogenetic studies two basic levels of repeating phenomena are often examined—one within a data set, and the other between data sets. Within a data set, congruent character evidence supporting monophyly of particular groups is one form of repetition, often providing well-supported nodes or branch points for cladograms. Between different data sets and analyses, parts or all of entire cladograms can repeat, providing taxonomic congruence.

Interpretation of patterns as evolutionary phenomena can be relatively theory minimal if evolutionary process assumptions used to generate those patterns are relatively minimal or absent. Most theories about evolutionary process are strongly influenced by individual beliefs rather than empirical evidence. It is suggested that in order for macroevolutionary studies to potentially discover predictive "constants" more equivalent to the "laws" or constants of the physical sciences (e.g., physics and chemistry), or for any fundamental improvement in the metatheoretical linkage between phylogenetic pattern and evolutionary process, we need to look more closely at character congruence and repeating natural patterns (including data from "taxonomic congruence") and, if possible, to depend less on belief and apriorism (e.g., metatheoretical influences).

Copyright © 1994 by Academic Press, Inc.
All rights of reproduction in any form reserved.

Introduction

REPEATING PATTERNS IN NATURE: To me, such patterns are the meat of science, the framework upon which scientific thought develops, and the major reason why science endeavors to develop process explanations. The use of patterns summarizing descriptive data to investigate historical processes is also an important distinction between science and religion. Scientists need not simply accept notions on faith, authoritarian dogma, or consensus. Scientists are usually skeptical and inquisitive. They attempt to base their theories on observations of nature (or descriptive data). Although scientists are also influenced by beliefs, they usually try to distinguish between personal belief and empirical observation. Before further discussion, I define my personal perspective.

My perspective is not that of a professional philosopher, but of a student of comparative morphology and biogeography trying to make historical sense of the diversity of fossil and living organisms. I am not interested in being identified as a pattern cladist or a transformed cladist, or even a traditional cladist. A cladistic pattern (cladogram) simply provides an efficient and relatively consistent way to organize empirical observations and look for order in nature. For any given pattern, congruence (a form of hierarchical compatibility) is used to hypothesize homology of descriptive features among taxa. One of my interests is in the occurrence of natural repeating patterns, whether individual patterns are based on morphologic, molecular, behavioral, or geographic features. By natural patterns, I mean nonrandom patterns that are not mere artifacts of statistics, taxonomy, or theory. Discovery of repeating cladistic patterns in nature, I believe, provides an excellent framework on which to propose and then to test process explanations. If such patterns in the biological sciences repeat enough that they become reliably predictable, they might eventually lead to something comparable to the predictive constants or "laws" of other sciences (e.g., gravity constant of physics).

This chapter focuses primarily on two issues. The first issue is the relative importance of theory minimal observations in evolutionary biology (followed by an example using transpacific patterns and the

Tertiary of western North America). The second issue is a brief review of the current debate between "total evidence" methods and "taxonomic congruence" methods of phylogenetic analysis.

THE IMPORTANCE OF THEORY MINIMAL OBSERVATIONS

Search for repeating patterns in nature can be relatively neutral with respect to evolutionary theory. As Brady (1982a, pp. 56-57, 1982b) and others have discussed, it is not possible to be absolutely theory-free in compiling observations. But even if it is impossible to be absolutely unbiased by preconceived theory in evolutionary biology, it is possible to be theory minimal by attaching as few process (e.g., Darwinian) explanations as possible to the search for repeating patterns in observed descriptive data. In a previous systematics volume, Hull (1988a, p. 35) wrote that "those scientists who have attempted to produce theory-free descriptions of natural phenomena have had pitifully little impact on the course of science". I disagree with his statement. If the development of a theory of continental drift, for example, was largely dependent on earlier discoveries of repeating patterns of relationship among organisms and/or magnetic anomalies (Wegener, 1912; Kummel, 1961, p. 343; Romer, 1968, p. 336; Rosen, 1981, p. 4; Menard, 1986), then discovery of patterns changed the course of science. Process explanations, such as sea-floor spreading and plate tectonics, moved things farther along that course as the patterns and process theories formed a more and more complicated synthesis. The development of the theory of evolution itself was partly dependent on earlier observations that groups of organisms have unique characterizations at different levels that fit congruently into hierarchical patterns. Many general phylogenetic patterns or "trees" accepted today are congruent with hierarchical patterns well established in pre-Darwinian time. Recognition of repeating patterns has been seminal to formulation of most significant concepts and theories of science.

Some repeating patterns may be easier to recognize than others. Repeating patterns in physics that led to formulation of a gravity constant were identified long ago because they were easy to observe.

When I release a pencil from the end of my desk, the pencil always falls to the floor. It also appears to take the same amount of time to reach the floor every time I drop it. Thus, a predictive component eventually arises from such observations. I can predict with a high degree of accuracy the time it will take for the pencil to hit the floor once I release it. This repeating pattern is so easy to observe that today it seems trivial (although it took over 1500 years of science to see the significance of it). Other repeating patterns, ones that could lead to analogous constants in systematics and macroevolutionary studies, have not yet been recognized, and perhaps this is because they are not as easy to observe as the falling pencil.

In comparative systematic studies of organisms, an investigator can start with the minimal hypothesis that it is possible to observe and identify structures in different organisms that are equivalent (e.g., topologically and functionally similar elements identified as potential homologies or synapomorphies). These elements (or characters) either show some degree of congruence, suggesting a specific hierarchical pattern (Fig. 1), or they vary randomly. Systematists look for patterns.

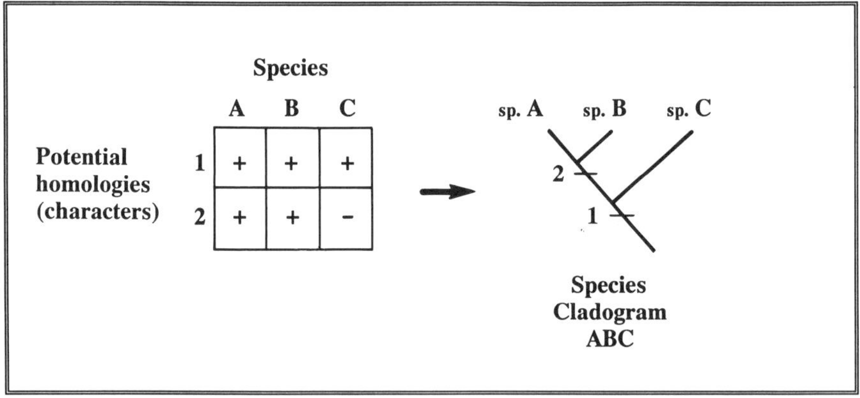

Figure 1. Summary of putative homologies (i.e., characters uniquely shared by groups of organisms) in a cladogram. Mostly after Grande and Micklich (1993).

But discovery of a single pattern (based on character congruence) could be thought of as a type of first-order descriptive information that could be used to find more general information in the form of *repeating* patterns (i.e., pattern congruence or "taxonomic congruence"). As stated by Nelson (1979, p. 8), "Replicated components are significant because the probability of replication for any given component, due to chance alone, is small". Kluge (1989) and others present arguments against the use of taxonomic congruence methods, and their points are discussed below in the section on "total evidence arguments".

If we look at skeletal features in a group of fishes, and find a well-supported pattern such as the one in Figure 1, we can then look at some other body of data for the same group of fishes, such as muscle anatomy, nerve anatomy, or molecular data. If a second set of data gives us the same pattern of relationship for taxa a, b, and c, then we have a repeating pattern, and a sort of second-order information. The more complex a repeating pattern is, and the more strongly (frequently) the same resolved pattern is repeated, the less likely it is that the pattern is due to random factors. There are a number of ways to search for second-order information of repeating patterns. One way (mentioned above) is to examine different systems, such as bones, muscles or molecules, independently. If a cladogram based on osteology produces a pattern of relationships that is the same as a pattern produced using muscle data or molecular data, then there is a repeating pattern. Another way to potentially discover repeating patterns is to examine geographic areas of endemism of the taxa being analyzed.

One way to define areas of endemism could be to define them as the putatively natural geographic ranges of species or other taxa (ignoring regions where the taxon is known to have been introduced by humans). If we take cladograms of taxa based on morphology and replace taxa with their areas of endemism, we can look for repeating geographic patterns. For example, if we have the two independent cladograms ABC and DEF in Figure 2, and we substitute the geographic ranges of the taxa for the species, we have a repeating pattern of area relationship (i.e., each group independently gives us the same area pattern). Areas one and two appear to be more closely related to each

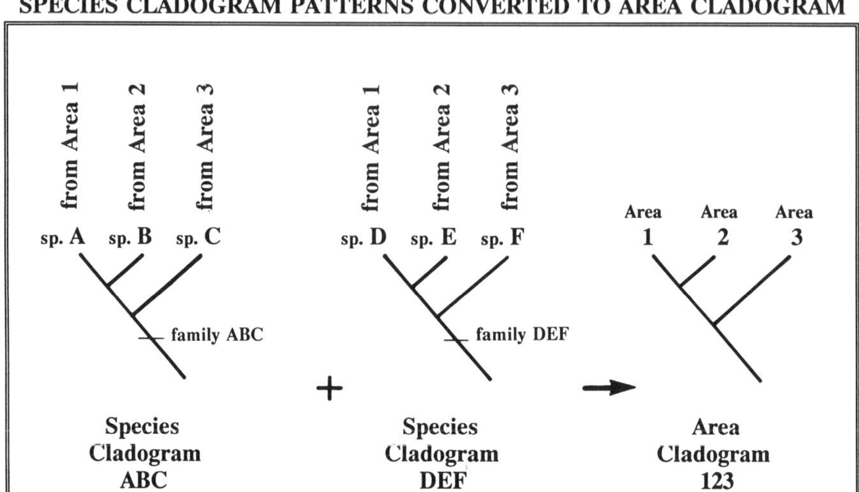

Figure 2. Conversion of species cladograms based on biological characters into an area cladogram. From Grande and Micklich (1993).

other than either is to area three. The more complex a well-supported area pattern is, and the more frequently an area pattern is found in other taxonomic groups, the less likely it is that the pattern is the result of random events. This technique of vicariance biogeography has been used or discussed at length in a number of other papers (e.g., Rosen, 1975, 1978; Platnick and Nelson, 1978; Nelson and Platnick, 1980, 1981; Wiley, 1981; Brown and Gibson, 1983; Grande, 1985, 1989). Once a repeating pattern of area relationships has been discovered, we can then look for process explanations by finding congruent patterns which summarize nonbiological events. For example, geological evidence indicating a two-step subdivision of a homogenous area 1+2+3 in Figure 3 produces an area pattern identical to the one produced from biological evidence in Figure 2.

AREA CLADOGRAM BASED ON GEOLOGICAL EVIDENCE

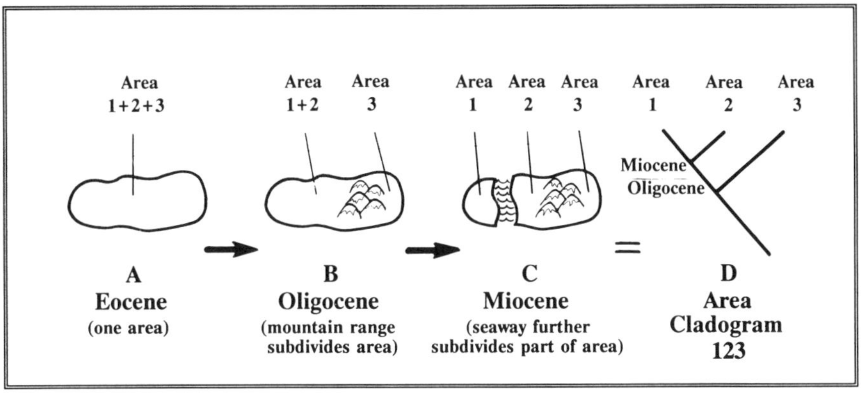

Area 1+2+3
A
Eocene
(one area)

Area 1+2 Area 3
B
Oligocene
(mountain range subdivides area)

Area 1 Area 2 Area 3
C
Miocene
(seaway further subdivides part of area)

Area 1 Area 2 Area 3
Miocene
Oligocene
D
Area Cladogram
123

Figure 3. An area cladogram based on nonbiological evidence (geological evidence in this example). In the Eocene, a single homogenous area exists which is divided into two areas in the Oligocene, and three areas in the Miocene. From Grande and Micklich (1993).

Once we have discovered a repeating pattern, our investigations can become more theory laden. As we look for possible process explanations for our pattern we often expand our theories with various scenarios and/or link our repeating pattern to other patterns, building a sort of synthesis of historical ideas. As we add layers of theory (e.g., drifting continents, natural selection, various species concepts) to our basic pattern information, our overall synthesis becomes less and less theory minimal. Also, as our investigations become more and more theory laden, "reciprocal illumination", a form of circular reasoning, can become more influential. Reciprocal illumination, as described by Hull (1988b, p. 132) occurs when "scientists reason from one sort of evidence to another, each time correcting past errors and expanding the scope of their hypotheses". Although reciprocal illumination consciously or subconsciously plays a major part in the way most or all evolutionary scientists formulate and refine their theories, a reasonable concern about reciprocal illumination is how much circular reasoning or manipulation we can incorporate into our data set before our new "discoveries" become merely empirically empty artifacts of our theories. I do not claim to have a quantitative answer to this question; but I believe that no scientific theory should be respected if it is not supported by observations that are logically independent of that theory.

Returning to the main topic of this chapter, the impact on science of theory minimal patterns observed in nature is examined. Below is an example of a study which starts with discovery of a repeating pattern, and becomes increasingly more theory laden as it develops. I review individual parts of this example and try to identify information that has scientific "impact".

TRANSPACIFIC PATTERNS AND THE TERTIARY OF WESTERN NORTH AMERICA

The Eocene Green River Formation represents an extinct great lake system–a complex biosystem whose demise seems correlated with major environmental change in western North America during the Tertiary period (Grande, 1989). Spanning about 15 million years or more (Grande, 1984) this was one of the world's longest lived lake systems, with a longevity exceeding even that of the Great African Rift lakes of today (Grande, in press). The Green River lake system formed in western North America, in parts of what are now Wyoming, Colorado, and Utah, during the late Paleocene, as the result of uplift of surrounding mountains, and downwarping of three major lake basins. Paleontological evidence indicates that during deposition of the main fossil beds the environment was subtropical fresh water (Grande, 1984, 1989).

Today much of the area is high mountain desert, where water is scarce and winters are long and cold. Starting in September, temperatures at night frequently fall below freezing. The present elevation of the Fossil Lake sediments is about 7500 ft above sea level. Consequently, we have an extinct tropical biota that can be analyzed for biogeographic information – a fresh set of historical data, in a sense

Table 1. Transoceanic affinities of the teleost groups that occur in the Green River Formation. Names with daggers (†) preceding them are extinct taxa. Reasons for unknown affinity are: *Reason A* - Species relationships unknown in sufficient detail to indicate transoceanic affinity. *Reason B* - Known family range is restricted to North America, and the family relationships are insufficiently known to indicate transoceanic affinity. *Reason C* - Known family range is restricted to North America, and the family is most closely related to a group with widespread distribution (including both transpacific and transatlantic). Thus no specific transoceanic relationship is specified.

Transoceanic relationships of Green River Formation teleosts		
Green River genus	Family that genus belongs in	Transoceanic affinity of Green River taxon
†*Eohiodon* (1 species)	Hiodontidae	transpacific (eastern Asia)
†*Phareodus* (2 species)	Osteoglossidae	transpacific (Australia and/or Indonesia)
†*Diplomystus* (1 species)	†Ellimmichthyidae	transpacific (China)
†*Knightia* (2 species)	Clupeidae (Subfamily Pellonulinae)	transpacific (China)
†*Gosiutichthys* (1 species)	Clupeidae (Subfamily Clupeinae)	unknown (reason A)
†*Notogoneus* (1 species)	Gonorynchidae	unknown (reason A)
†*Amyzon* (1 species)	Catostomidae	transpacific (China)
†*Astephus* (1 species)	Ictaluridae	unknown (reason B)
†*Hypsidoris* (1 species)	†Hypsidoridae	unknown (reason C)
†*Erismatopterus* (1 species)	Percopsidae	unknown (reason B)
†*Amphiplaga* (1 species)	Percopsidae	unknown (reason B)
†*Asineops* (1 species)	†Asineopidae	unknown (reasons A and B)
†*Mioplosus* (1 species)	unknown	unknown (reason A)
†"*Priscacara*" (currently a form genus only. Several Green River species currently contained in this genus.)	unknown	unknown (reason A for all species)

independent of the extant biota.

Fossilized organisms of the Green River deposits, particularly fishes, are so abundant and so well preserved that they can be included in detailed phylogenetic studies with extant and other well-preserved fossil taxa. We can use the comparative anatomy of Green River organisms phylogenetically with species from elsewhere, to search for repeating patterns relevant to earth history.

Elsewhere (Grande, 1985, 1989, in press; Grande and Bemis, 1991) I have summarized available phylogenetic information for Green River teleost fishes and noted a repeating pattern of area relationship between early Tertiary western North America and the western Pacific region. When I first inferred a transpacific pattern for Green River teleosts (Grande, 1985, Table 1; 1989, Table 1) I found that five of six groups independently show transpacific relationships. Only the Percopsidae, based on a study by Rosen and Patterson (1969), appeared to show a transatlantic pattern. Later, Patterson and Rosen (1989) revised their pattern of relationships for Percopsidae, based on additional morphological data, and the Green River percopsids no longer indicate a transatlantic relationship. Consequently, all Green River teleosts that presently show transoceanic relationships for North America, show transpacific relationship (Grande and Bemis, 1991, p. 115). This information is summarized in Table 1. Five taxa each independently show transpacific relationship (based on various studies cited in Grande, 1985); and none shows a transatlantic one. Using all phylogenies currently available for the Green River species, it seems significant that a wide variety of ichthyologists, studying a wide variety of families, discovered the same transoceanic pattern.

I did not expect *a priori* to find the transpacific pattern shown by this fauna because the extant fauna of western North America does not show it (in part due to absence of Hiodontidae, Osteoglossidae, Ellimmichthyidae, and Pellonulinae from the extant western North American fish fauna). Discovery of this initial repeating pattern represents the least theory-laden part of this investigation. Once we start to search for corroboration from other groups of organisms, or nonbiological information, the study becomes more theory-laden. For example, if we scan work on other groups of Green River actinopterygian fishes, we find that the pattern shown by the Green River teleost fauna is further repeated by the paddlefish *Crossopholis*. The closest known transoceanic relative of *Crossopholis* from the Green River Formation is the living *Psephurus*, from China. But paddlefishes and teleosts do not form a monophyletic group, and the groups phylogenetically in between (gars and bowfins, see Fig. 4) do not show congruent area patterns. The gars from the Green River Formation, according to a phylogeny by Wiley (1976), indicate a transatlantic relationship rather than a transpacific one. Therefore we have no strong reason to suspect a transpacific relationship for any group of non-teleostean fishes, because gars, which are more closely related to teleosts than to paddlefishes (Fig. 4), show a transatlantic relationship.

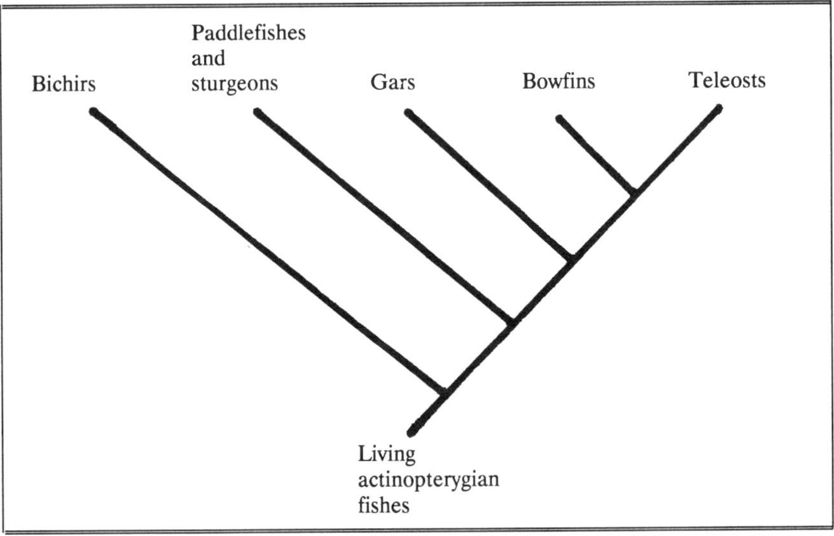

Figure 4. A cladogram of living actinopterygian fishes, after Patterson and Rosen (1977) and Rosen et al. (1981).

Based on their phylogenetic relationships (Fig. 4) gars and paddlefishes are relatively older groups than teleosts, and may possibly predate the causal mechanism for the teleost patterns (as would placoderms, acanthodians, paleoniscoids, etc.). Within one monophyletic group (teleosts) a transpacific pattern is indicated by five out of five independent lineages. Within another monophyletic group (paddlefishes and sturgeons) only one indicates a transpacific pattern. We can theorize that paddlefishes, showing transpacific ties, may have been affected by the same process that resulted in the repeating transpacific pattern for teleosts, but because there is only one paddlefish phylogeny indicating transpacific relationship, the pattern is nonrepeating. Because it is nonrepeating, we have no statistical reason to believe it nonrandom (e.g., it could be the result of extinction of a transatlantic species). It is the statistical equivalent of producing a resolved three-taxon cladogram using a single character (i.e., we need some test of congruence to polarize a single character; many congruent characters with little or no homoplasy indicate nonrandom order more strongly). When Grande and Bemis (1991, p. 115) chose to include the area pattern indicated by paddlefishes as part of the transpacific pattern shown by teleosts, it was proposed as a theory unsupported by repeating patterns. Only if other groups most closely related to paddlefishes also showed a transpacific pattern (e.g., sturgeons) would paddlefish data become part of another repeating pattern (i.e., within Acipenseriformes).

As mentioned above, the strong repeating pattern that we find for Eocene teleosts of western North America has not been demonstrated for the area today with the extant teleost fauna, possibly indicating that the biogeographic pattern for the area has changed through time. (More likely it may indicate that no one has critically tried to analyze the transoceanic affinity of the living western North American fish fauna.) A changing pattern of area relationship for western North America has been inferred by MacGinitie in a 1969 monograph on the Green River plants. In summarizing the flora he wrote: "representatives of living Asiatic species in the earlier Tertiary floras of the Western states, are typical" and "the composition of Western fossil floras was more Asiatic than European, until late in the Tertiary." It was not until the late Cenozoic, that he noted an influx of European forms. So there are two major phenomena for western North America indicated by fossil fishes and/or plants. One is a strong transpacific pattern of relationship for Eocene western North America, and a second is a later loss of that pattern, due to extinction and also possibly due to mixing of the transpacific biota with European forms later in the Tertiary, presumably from eastern North America and a North Atlantic connection.

How do these events correlate with the presumed geological history of North America? It is reasonable to assume that early Cenozoic barriers such as the Upper Cretaceous seaway, and mountain building activity in the Rocky Mountain region isolated major components of the terrestrial biotas of eastern and western North America from each other for a long time. Temporary removal of major barriers in the Oligocene, when the Rockies were thought to have been eroded to their lowest point (Scott, 1975), may have been a time when the biotas of eastern and western North America underwent a period of mixing. I theorize that the original western North American teleost fauna was transpacific in origin, even though the one today may be more of a hybrid fauna containing transatlantic elements also.

Based on this theory of origin and history I predict that if we ever find a well preserved Paleocene or Eocene freshwater teleost fauna from eastern North America, its biogeographic affinities will be European, even though the extant fauna is a mix of European and Asian. Also I predict that we will see similar biogeographic patterns turn up in other groups of Eocene organisms, once the phylogenetic relationships of the many yet undescribed Green River Formation birds, lizards, and other vertebrates have been determined.

What explains the origin of a transpacific biota in western North America (including Australian components)? A number of authors have discussed the probable exotic origin of parts of the North American west coast land mass (Craw and Weston, 1984; Coney et al., 1980; Kerr, 1980; McWilliams and Howell, 1982; and others). One controversial theory that would explain a transpacific biota is the Pacifica hypothesis of Nur and Ben-Avraham (1977, 1978, 1981), geophysicists who proposed a land mass, which they called Pacifica, that was initially connected to Australia, as shown in Figure 5. Initially Pacifica was a single continent as late as the Permian (Fig. 5A). During the Jurassic and Cretaceous (Fig. 5B-D), the land mass broke up where a triple-junction, sea floor spreading ridge exists today. Pieces of this land mass were dispersed by sea floor spreading and incorporated into eastern Asia, western North America, and western South America, where exotic geological terranes are found today. A North American block, Wrangellia, which has been known for a long time, is thought to have collided with western North America during the middle or late Cretaceous, a time when we see also a change in the flora and fauna of North America. It is possible that the transpacific fauna is actually descended from a Pacifica fauna, dispersed around the Pacific rim by rafted pieces of Pacifica. There are arguments against the Pacifica hypothesis as a method of transporting organisms across the Pacific (Tedford, 1981; McKenna, 1981; Batten and Schweickert, 1981) but it is, nevertheless, a possible explanation.

REPEATING PATTERNS AND "IMPACT" IN SCIENCE

If I correctly interpret Hull's (1988a, and above) quote that scientists who attempt to produce theory-free descriptions of natural history phenomena have pitifully little impact, I see the implication that more theory-laden concepts have more impact on science. While the term "impact" could be somewhat controversial, we can examine components of the Green River Formation study presented above.

What part of this study could potentially have the most impact on our scientific knowledge about earth history and evolution? Everything is relative, and never mind the possible importance or lack of importance of the specific study described above (there is always the possibility that my observations could be wrong). Let us look at the individual parts of this study, and assume no other information. One way to compare the relative importance of various components of this project is to examine the effect of removing various components of information. First, there was the discovery of the transpacific pattern for an area of western North America. Although certainly theory laden,

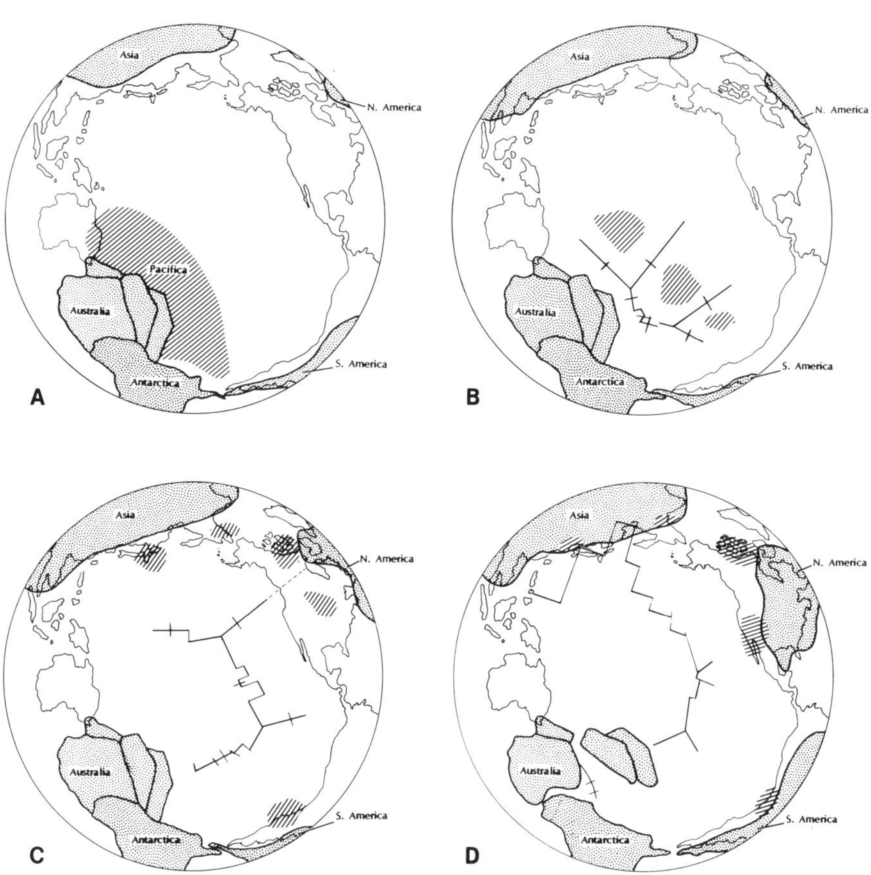

Figure 5. Schematic model after Nur and Ben-Avraham (1981) of their proposed breakup of Pacifica and resulting collision events. (A) 225 million years before present (MYBP), (B) 180 MYBP, (C) 135 MYBP, and (D) 65 MYBP. Fine lines mark present-day continental outline. Heavy lines mark locations of various continental areas (stipple) through time (after Dietz and Holden, 1970). Position of spreading centers simplified from Larson and Chase (1972), Uyeda and Ben-Avraham (1972), Ben-Avraham and Uyeda (1973), Hilde et al. (1976), and Hayes and Ringis (1973). Adapted from Nur and Ben-Avraham (1981).

this is the least theory-laden part of the whole scenario. This was followed by a series of more and more theory-laden bits of historical information.

If we remove the Pacifica hypothesis of Nur and Ben-Avraham, what happens? We are left to choose another causal explanation for the repeating pattern of area relationships, and other parts of the scenario remain intact. If we remove the corroborating information about fossil plants or paddlefishes, we still have various teleost groups indicating a strong area pattern, and the rest of the scenario remains intact. If we remove the scenario about mountain ranges and seaways providing barriers to organisms, we could come up with an ecological barrier or some other type of barrier instead, but we are still left with a repeating pattern of area relationship and the rest of the scenario. Even if we remove some teleost fish groups, including some that show the transpacific area pattern, we do not change anything significant about the whole picture. But if we remove the initial transpacific area pattern for western North America, everything collapses. There is no repeating (i.e., putatively nonrandom) pattern from which to extrapolate and for which to build complex scenarios. The original repeating pattern is also the most objective part of the research project because it is least affected by "reciprocal illumination". Only later might it be churned and recycled, in a number of forms, to market to the scientific community.

TOTAL EVIDENCE ARGUMENTS AGAINST USE OF CERTAIN REPEATING PATTERNS

A number of contemporary systematists (e.g., Jones, Kluge and Wolf, 1993; Donoghue, Doyle, Gauthier, Kluge and Rowe, 1989) have followed Kluge (e.g., 1989) in arguing for the use of the so-called "total evidence" method of analysis (combining all available data into a single cladogram). This method eliminates what Kluge (e.g., 1989) refers to as "taxonomic congruence" (congruence between cladograms based on independent data sets) and thereby eliminates the possibility of detecting pattern congruence from independent data sets. Reasons given by the "total evidence" school against using taxonomic congruence methods fall primarily into two categories: (1) *Subdivisions of data are artificial* (e.g.,

Kluge and Wolf, 1993, state that "such subdivisions [of data] do not provide a reasonable basis for claiming the existence of classes of evidence"). (2) *Taxonomic congruence methods fail to resolve problems of incongruence between conflicting cladograms* (e.g., Jones et al., 1993, p. 93, state that "[with taxonomic congruence] the more fundamental cladograms differ, the more the consensus hypothesis will be unresolved and the greater will be the loss of descriptive and explanatory power").

My response to the first reason against use of taxonomic congruence is that data can be subdivided in a number of ways for practical and theoretical reasons. Bone specialists tend to look at bones, muscle specialists at muscles, and so on; or with regard to vicariance biogeography, catfish specialists study catfishes, beetle specialists study beetles. We can each qualitatively evaluate some systems or taxa better than others, because no specialist is equally competent at all things (e.g., organ systems, molecules, or taxa). Subdivision of data is a fact of life that we have to deal with. Also, we will never have a "complete" data set (or even know how to identify such a thing), because some taxa have effectively an infinite number of potential characters, while others (extinct taxa) will never be known, or will be known incompletely. A larger data set from the "total evidence" method may give us a different cladistic pattern than independent or other partial data sets, but can we really say whether it is better or worse than the other patterns? Donoghue et al. (1989, p. 456) claim that by not using the total evidence method (leaving out data from fossil taxa in their case) "in some cases...the true phylogeny cannot be obtained". How do we identify truth in science? I believe that, as with assessing the utility of characters on a cladogram, the utility of any data set, and the utility of subdividing data sets, can be evaluated only by congruence.

My response to reason 2 against use of taxonomic congruence is that it misses the main point of taxonomic congruence (and the search for repeating patterns). I believe that most (or at least many) systematists interested in taxonomic congruence are searching for multiple repeating patterns (i.e., congruence or pattern homology) and not simply trying to resolve conflicts between two incongruent cladograms or classifications. At least with regard to finding a link between phylogenetic pattern and evolutionary processes, there is a hope that historical processes have left a strong enough signal in the form of multiple repeating patterns that it will be identifiable against a background of noise (incongruent cladograms, homoplasy resulting from human error or inadequate methodology). If, after decades of work, the best that phylogenetic studies had done was to produce conflicting, nearly equally parsimonious cladograms, then I believe evolutionary studies would be in serious trouble. I do not think this is the case. Repeating patterns do exist, and even in those cases in which conflicting, nearly equally parsimonious cladograms exist, there are at least one or more nodes or branch points that repeat. Perhaps those are nodes that we should be most concerned with in looking at the broadest questions about linking phylogenetic pattern to evolutionary process (e.g., determining what metatheories of biology, *sensu* Nelson, 1989, are relevant to phylogenetic studies). I believe that the most significant theories in natural history have come from those cases in which there has been pattern congruence between different sets of data (e.g., theories concerning moving continents, or the very notion of evolution itself developed from observations of repeating patterns). In some cases I see a potential "loss of descriptive and explanatory power" (quoted above) resulting from using only total evidence methods, and ignoring potential information from taxonomic congruence methods. For example, if relatively limited amounts of osteological data, neurological data, myological data and behavioral data each produce identical cladograms (i.e., show a repeating pattern), but an overwhelming amount of data from molecular analysis of a particular protein produces a conflicting cladogram, the repeating pattern signal could be obscured. Without segregation of data into the original subsets, we may be unable to identify the incongruence as exclusive to a single type of data or to a single carelessly done study. Since we have no way of demonstrating that all characters are equivalent (any more than we can demonstrate that they are not), we should not allow a single conflicting pattern from one type of data to obscure a pattern repeated by several other types of data. That is not to say that the total evidence pattern is wrong, but rather that we cannot afford to dismiss the repeating pattern as being less likely to reveal a historical signal. Any pattern obtained through congruence is a potential tool in pattern/process studies.

Part of the reason for disagreement between the total evidence school and viewpoints supporting taxonomic congruence is in the nature of arguments presented. The total evidence argument is strictly quantitative, whereas the taxonomic congruence argument is also qualitative. The issue of integrating pattern with process is both a quantitative issue and a qualitative philosophical issue, especially if terms like "true phylogeny" (quote from Donoghue et al., above) are used. Limiting ourselves to a single, quantitative approach to the problem of matching evolutionary process to pattern implies that we know the best equation. I contend we do not know what "best" is in this case. Consequently, total evidence and taxonomic congruence methods should both continue to be applied together in evolutionary/ phylogenetic studies.

Conclusion

Today a successful scientific career in evolutionary biology often seems to require few original or fundamental discoveries and much creative scenario building based more on other theories than on descriptive data. I don't want to speculate on whether this is good or bad, but I believe that without discovery of more repeating patterns in systematics, those of us interested in evolutionary process and its connection to phylogenetic pattern will be stuck in a rut. If we are ever to discover evolutionary constants comparable to the so-called laws of physics and chemistry, I believe they will first be indicated by widespread repeating patterns. Of course it is possible that we may never discover such constants in biology, and a smoother synthesis of pattern and process will never be made, but I remain an optimist. I also unambiguously identify my optimism as a result of my beliefs rather than of scientific evidence. Distinguishing between one's beliefs and one's scientific evidence may be one of the most difficult aspects of evolutionary studies.

I hope that there are yet undiscovered repeating patterns in nature that will be identified and result in new concepts of major importance to evolutionary biology. These new patterns may result from new data, or possibly from new ways of analyzing existing data. Without discovery of new repeating patterns, studies of evolutionary process as applied to phylogeny will continue to be driven largely by belief and popular consensus rather than observation of nature. Like different houses of cards, some theories will be based on other theories that themselves are contingent upon other theories. In any case, impact and importance of contemporary science seem subjective concepts at best, and true impact and importance will be determined only in retrospect.

Acknowledgments

For discussions and/or reading and commenting on various versions of this manuscript I thank Gareth Nelson, Colin Patterson, Mario de Pinna, Olivier Rieppel, Joel Cracraft, and Paulo Buckup. Their comments improved the sense and style of this paper. I also thank Olivier Rieppel for originally organizing the interesting and provocative symposium that led to this volume. Some support for this paper came from NSF BSR 9119561.

References

Batten, R. L., and R. A. Schweickert. 1981. Discussion [a reply to Nur and Ben-Avraham, 1981]. In G. Nelson and D. E. Rosen (Eds.), *Vicariance Biogeography: A Critique.* New York: Columbia University Press. 593 pp.

Ben-Avraham, Z., and S. Uyeda. 1973. The evolution of the China Basin and the Mesozoic paleogeography of Borneo. *Earth and*

Planetary Science Letters, 18(2): 365-376.

Brady, R. H. 1982a. Parsimony, hierarchy, and biological implications. In N. I. Platnick and V. A. Fink (Eds.), *Advances in Cladistics*, Vol. 2, 49-60. Proceedings of the Second Meeting of the Willi Hennig Society. New York: Columbia University Press.

Brady, R. H. 1982b. Theoretical issues and "pattern cladistics". *Systematic Zoology*, 31: 286-291.

Brown, J. H., and A. C. Gibson. 1983. *Biogeography*. St. Louis, MO: C. V. Mosby Co.

Coney, P. J., D. L. Jones and J. W. H. Monger. 1980. Cordilleran suspect terranes. *Nature (London)*, 288: 329-333.

Craw, R. C., and P. Weston. 1984. Panbiogeography: a progressive research program? *Systematic Zoology*, 33(1): 1-13.

Dietz, R. S., and J. C. Holden. 1970. Reconstruction of Pangaea: breakup and dispersion of continents, Permian to present. *Journal of Geophysical Research*, 75(26): 4939-4956.

Donoghue, M. J., J. A. Doyle, J. Gauthier, A. G. Kluge and T. Rowe. 1989. The importance of fossils in phylogenetic reconstruction. *Annual Review of Ecology and Systematics*, 20: 431-460.

Grande, L. 1984. Paleontology of the Green River Formation, with a review of the fish fauna. Second edition. *Geological Survey of Wyoming, Bulletin*, 63: 1-333.

Grande, L. 1985. The use of paleontology in systematics and biogeography, and a time control refinement for historical biogeography. *Paleobiology*, 11(2): 1-11.

Grande, L. 1989. The Eocene Green River lake system, Fossil Lake, and the history of the North American fish fauna. In J. Flynn (Ed.), *Mesozoic/Cenozoic Vertebrate Paleontology: Classic Localities, Contemporary Approaches*, 18-28. 28th International Geological Congress Fieldtrip Guidebook T322. Washington, DC: American Geophysical Union.

Grande, L. (in press). The Tertiary Green River Lake Complex, with comments on paleoenvironments and historical biogeography. *Contributions to Geology*.

Grande, L., and W. Bemis. 1991. Osteology and phylogenetic

relationships of fossil and Recent paddlefishes (Polyodontidae) with comments on the interrelationships of Acipenseriformes: *Society of Vertebrate Paleontology, Memoir 1*, i-viii, 1-121, supplement to Journal of Vertebrate Paleontology, 11(1).

Grande, L., and N. Micklich. 1993. Paleobiogeography of the Eocene Messel and Geiseltal fish faunas. In *Gruße Messel - Perspectives and Relationships. Kaupia*, 3(2): 243-255.

Hayes, D. E., and J. Ringis. 1973. Seafloor spreading in the Tasman Sea. *Nature (London)*, 243(5408): 454-458.

Hilde, T. W. C., S. Uyeda and L. Kroenke. 1976. Evolution of the western Pacific and its margin. In C. L. Drake (Ed.), *Geodynamics: Progress and Prospects*, 1-15. Washington, DC: American Geophysical Union.

Hull, D. 1988a. Progress in ideas of progress. In M. Nitecki (Ed.), *Evolutionary Progress*, 27-48. Chicago: University of Chicago Press.

Hull, D. 1988b. *Science as a Process. An Evolutionary Account of the Social and Conceptual Development of Science.* Chicago: University of Chicago Press.

Jones, T. R., A. G. Kluge and A. J. Wolf. 1993. When theories and methodologies clash: a phylogenetic reanalysis of the North American ambystomatid salamanders (Caudata: Ambystomatidae). *Systematic Biology*, 42(1): 92-102.

Kerr, R. A. 1980. The bits and pieces of plate tectonics. *Science*, 207: 1059-1061.

Kluge, A. G. 1989. A concern for evidence and a phylogenetic hypothesis of relationships among *Epicrates* (Boidae, Serpentes). *Systematic Zoology*, 38(1): 7-25.

Kluge, A. G., and A. J. Wolf. 1993. Cladistics: what's in a word? *Cladistics*, 9(2): 183-199.

Kummel, B. 1961. *History of the Earth*. W. H. Freeman and Company. 610 pp.

Larson, R. L., and C. G. Chase. 1972. Late Mesozoic evolution of the western Pacific Ocean. *Bulletin of the Geological Society of America*, 83(12): 3627-3644.

McKenna, M. 1981. Discussion [a reply to A. Hallam, 1981]. In G. Nelson and D. E. Rosen (Eds.), *Vicariance Biogeography: A Critique*, 335-338. New York: Columbia University Press. 593 pp.

McWilliams, M. O., and D. G. Howell. 1982. Exotic terranes of western California. *Nature (London)*, 297: 215-217.

MacGinitie, H. D. 1969. The Eocene Green River flora of northwestern Colorado and northeastern Utah. *University of California Publications in the Geological Sciences*, 83. 140 pp.

Menard, H. W. 1986. *The Ocean of Truth: A Personal History of Global Tectonics*. Princeton University Press.

Nelson, G. 1979. Cladistic analysis and synthesis: principles and definitions, with a historical note on Adanson's *Familles des plantes* (1763-1764). *Systematic Zoology*, 28: 1-21.

Nelson, G. 1989. Cladistics and evolutionary models. *Cladistics*, 5: 275-289.

Nelson, G., and N. I. Platnick. 1980. A vicariance approach to historical biogeography. *BioScience*, 30(5): 339-343.

Nelson, G., and N. I. Platnick. 1981. *Systematics and Biogeography: Cladistics and Vicariance*. New York: Columbia University Press. 567 pp.

Nur, A., and Z. Ben-Avraham. 1977. Lost Pacifica continent. *Nature (London)*, 270(5632): 41-43.

Nur, A., and Z. Ben-Avraham. 1978. Speculations on mountain building and the lost Pacifica continent. *Journal of the Physics of the Earth, Supplement*, 26: 21-37.

Nur, A., and Z. Ben-Avraham. 1981. Lost Pacifica continent: a mobilistic speculation. In G. Nelson and D. E. Rosen (Eds.), *Vicariance Biogeography: A Critique*, 341-358. New York: Columbia University Press. 593 pp.

Patterson, C., and D. E. Rosen. 1977. A review of ichthyodectiform and other Mesozoic teleost fishes, and the theory and practice of classifying fossils. *Bulletin of the American Museum of Natural History*, 158(art.2): 81-172.

Patterson, C., and D. E. Rosen. 1989. The Paracanthopterygii revisited: order and disorder. In D. Cohen (Ed.), *Papers on the Systematics of Gadiform Fishes*, 5-36. Natural History Museum of Los

Angeles County, Science Series, no. 32.

Platnick, N. I., and G. Nelson. 1978. A method of analysis for historical biogeography. *Systematic Zoology*, 27: 1-16.

Romer, A. S. 1968. Fossils and Gondwanaland. *Proceedings of the American Philosophical Society*, 112(5): 335-343.

Rosen, D. E. 1975. A vicariance model of Caribbean biogeography. *Systematic Zoology*, 24: 431-464.

Rosen, D. E. 1978. Vicariant patterns and historical explanation in biogeography. *Systematic Zoology*, 27: 159-188.

Rosen, D. E. 1981. Introduction [to *Vicariance Biogeography: A Critique*]. In G. Nelson and D. E. Rosen (Eds.), *Vicariance Biogeography: A Critique*, 1-5. New York: Columbia University Press. 593 pp.

Rosen, D. E., and C. Patterson. 1969. The structure and relationships of the paracanthopterygian fishes. *Bulletin of the American Museum of Natural History*, 141: 357-474.

Rosen, D. E., P. L. Forey, B. G. Gardiner and C. Patterson. 1981. Lungfishes, tetrapods, paleontology and plesiomorphy. *Bulletin of the American Museum of Natural History*, 167: 159-276.

Scott, G. R. 1975. Cenozoic surfaces and deposits in the southern Rocky Mountains. *Geological Society of America Memoir*, 141: 227-248.

Tedford, R. H. 1981. Discussion [a reply to Nur and Ben-Avraham, 1981]. In G. Nelson and D. E. Rosen (Eds.), *Vicariance Biogeography: A Critique*, 367-370. New York: Columbia University Press. 593 pp.

Uyeda, S., and Z. Ben-Avraham. 1972. Origin and development of the Philippine Sea. *Nature Physical Science*, 240: 176-178.

Wegener, A. 1912. Die Entstehung der Kontinente. *Geologische Rundschau*, 3(4): 276-292.

Wiley, E. O. 1976. The phylogeny and biogeography of fossil and Recent gars (Actinopterygii: Lepisosteidae). *University of Kansas Museum of Natural History, Miscellaneous Publication*, 64: 1-111.

Wiley, E. O. 1981. *Phylogenetics: The Theory and Practice of Phylogenetic Systematics*. New York: John Wiley and Sons. 439 pp.

5

Morphological and Molecular Inroads to Phylogeny

Michael J. Novacek

Department of Vertebrate Paleontology
American Museum of Natural History
New York, New York 10024

Abstract. Studies of molecular and morphological data share the same assumptions regarding homology and the same mission in construction of cladograms indicative of phylogenies. Both approaches are also tied by the same analytical methods. In a sense, molecular and morphological data are part of one broad spectrum of evidence bearing on structure and relationships. Yet morphological and molecular systematics involve somewhat different operations. The former relies on homology statements for diverse taxa and is enriched by ontogenetic information and the fossil record. The latter has enormous potential for massive character information that is directly related to heritable configurations in organisms. Particular problems, such as uninformative properties of certain genes or certain categories of substitutions and the incomplete preservation of fossils, also confound either approach. Development of several examples, with particular emphasis on the problem of bat origins and affinities, reveals some of the benefits of applying varying weighting schemes, adding or removing test taxa, constraining data based on independent evidence, evaluating tree length distributions, and combining data sets.

Copyright © 1994 by Academic Press, Inc.
All rights of reproduction in any form reserved.

Introduction

MORPHOLOGICAL AND MOLECULAR STUDIES OFFER STRIKINGLY different ways to uncover the history of life. One approach is very old, the other virtually newborn in comparison. One approach deals broadly with structure, the other with parcels of structure too minuscule for direct observation. One approach draws on very complex and interpretive descriptions, the other on a depauperate alphabet used as a code for infinite variations on a four-note theme. The potentials of the two approaches also differ distinctly (Hillis and Dixon, 1989). The greatest advantage of molecular data is that they can potentially retrieve all the heritable information. The morphological data set in its entirety is only some unknown fraction of this heritable information. On the other hand, molecular data apply to only a fraction of all known organisms. It is unlikely that the vast array of phylogenetic information in fossils–information bearing on the relationships of both fossils and extant taxa (Gauthier et al., 1988; Donoghue et al., 1989; Novacek, 1992)– will ever be accessed through molecular studies. This limitation holds despite the remarkable assays of gene sequences in very young and extraordinarily preserved fossils (Golenberg et al., 1990; DeSalle et al., 1992).

Yet any contrast between morphological and molecular data should not minimize their similarities as phylogenetic evidence. Both are applicable to the same methods of analysis (parsimony, distance, maximum likelihood, etc.). Both are weakened by faulty homology statements or poor sampling of characters and taxa. And both can, in some instances, produce identical phylogenetic patterns (hereafter referred to as trees) for the same taxa. Accordingly, comparisons between molecular and morphological approaches have been made before (Patterson, 1987; Hillis, 1987; Wyss et al., 1987; Novacek et al., 1988; Goodman, 1989; Swofford, 1991a) and will likely be made again. There is now an explosive increase of DNA sequence information. As sampling improves in the molecular area, so will our basis for comparisons with morphological work. This is an interim assessment that highlights a few of the issues.

In this chapter, I wish to consider molecular and morphological approaches under the basic operations common to them; namely, homology statements, generating phylogenetic patterns for tree construction, sampling strategies, and evaluation of results. Coverage is not tantamount to a recipe for phylogenetic analysis. There are texts and manuals available for that purpose. Discussion is instead slanted toward four questions. What do we uniquely gain with either molecular or morphological approaches? Conversely, are there significant liabilities in either method? How are molecular and morphological results compared and evaluated? Is it feasible to combine morphological and molecular data in phylogenetic reconstruction? Reference to case studies will highlight the current investigations of bat origins and relationships. This issue is the focus of a spate of recent molecular and morphological work.

Discussion here is largely confined to morphological trait data and protein and nucleotide sequence data. This does not deny the value of other kinds of evidence (e.g., immunological or DNA hybridization). A phylogenetic signal could potentially be picked up from several different sources. I justify, however, the scope herein with a simple argument. Homology, the central issue in phylogenetics, is only accessible from studies of explicit characters. If distance data tell us that two taxa are close, they say only that. Further tests within this framework merely demonstrate whether another taxon is closer or another protein or set of measurements changes the distance between the taxa. Beyond this series of tests based on increased sampling, we gain little knowledge about the nature of the evidence. In contrast, explicit morphological characters and molecular sequences considerably enrich our notion of homology; and homology is what constitutes the evidence for relationships. It is not surprising that the surge of activity in systematics concerns the study of homology and relationships based on explicit traits.

Remarks here on the nonmolecular side refer only to the evidence of morphology. Of course, the basic qualities of morphological data extend to any observations of the phenotype, for example, physiology and behavior. Homologies in structures (for example, the wings of bats or birds) can combine with behavioral or physiological attributes that together provide evidence of relationship.

HOMOLOGY STATEMENTS

Is there an essential difference between molecular and morphological approaches with respect to the problem of homology? Patterson's (1989, p. 473) remarks are illuminating here:

"So in terms of method, I believe that morphologists have been engrossed in the search for synapomorphy, or with probing their data; whereas molecular systematists have been engrossed with how to get the data (sequencing and other techniques) and second how to process them (algorithms). For the morphological systematist, an unerring guide to synapomorphy is at rainbow's end; for those of us who equate synapomorphy and homology, our search is for the key to homology. It follows that morphologists using PAUP (for instance) expect the algorithm to tell us which correspondences between taxa are homologous, or synapomorphous, or characterize natural groups. The molecular systematist's approach is somewhat different, and the difference is illuminated by the frequency with which homology and similarity are equated in the molecular literature [Reeck et al., 1987], and by Dover's comment [1987] that "homology and similarity are operationally synonymous." In molecular systematics, homology tends to be treated as a given, an *a priori* assumption...."

I take issue with this distinction. Homology at some level must be treated as a given–"an *a priori* assumption"–in any morphological analysis. The data matrix from which a favored pattern of synapomorphy is eventually revealed reflects a set of assumptions concerning homology in a broader context. Homology at this level boils down to the recognition of similarities broadly distributed among taxa. The discovery of homology (equals synapomorphy) through a parsimony analysis of the matrix simply represents a refinement dependent upon the system assumed. We cannot make a meaningful homology statement about wings of all bats without assuming they derive from forelimbs shared by all tetrapods.

At the same time, it is increasingly evident that the equation of similarity and homology in molecular studies is far from trivial. Matching (alignment) of sequences in different taxa (orthologous sequences) is confuted in many cases by ambiguities resulting from insertion and deletion (gap) events (Waterman et al., 1991). These gap events are, in themselves, potential indications of homologies. Although alignments are relatively easy in protein-coding genes they are far from so in nonprotein-coding sequences, such as rRNA or mitochondrial D-loop regions. This has prompted a reconsideration of the mode for alignment–in a basic search for similarities–that conforms more closely to Patterson's characterization of the morphologist's search for homology. Thus, Mindell (1991, p. 73, 74) maintains "...species sequences should be aligned in the descending order of phylogenetic relationships (phylogenetic weighting of alignments) to maintain the continuity of information which forms the basis of relationships of homology." A somewhat different interpretation of alignment, but one also linked to synapomorphy and homology, is offered by Wheeler and Gladstein (1992); they note the minimum number of steps required by an alignment is naturally the most parsimonious branching diagram (cladogram) for these sequences. It is therefore logical, Wheeler and Gladstein argue, to extend this criterion to the alignments themselves, a mode that underscores their program for multiple alignment (MALIGN©).

This shared emphasis on homology in molecular and morphological work can be illustrated by the application of parsimony in both approaches. In the former, aligned sequences are not recoded as primitive ("0") or derived ("1, 2...n") states (exceptions include a recoding by Wyss et al., 1987, of amino acid replacements for several proteins). How then are synapomorphies (aka homologies) for these nucleotide sequences recognized? Customarily, the data set includes one or more outgroup reference taxa. Rarely, if ever, does this outgroup provide the ideal ancestral sequence for the remaining taxa; not all of its nucleotide sequences represent the original states for all taxa in the data set. Parsimony applied to both the ingroup and the outgroup taxa (Maddison et al., 1984) may dictate that some of the loci in the outgroup taxon actually show departures from the likely ancestral sequence. Likewise, morphological parsimony applications involve outgroup taxa that are not necessarily ideal ancestors for the remaining taxa. The *a priori* recognition of polarity (primitive or ancestral versus derived states) in morphology as a method somehow detached from outgroup plus ingroup parsimony is unwarranted (Maddison et al., 1984; Swofford, 1991b). A similarity in structure at a broader level in the reference system (e.g., forelimbs in tetrapods) can be distinguished from similarities that are much more localized (the featherless wings of bats). Parsimony analysis will further provide evidence relevant to a decision whether or not these localized similarities are homologous in all the taxa that share them.

Despite this basic conformity, morphological and molecular studies do confront different operational problems. A data matrix of morphological characters is customarily enriched by information on the hierarchical distribution of characters in a broad diversity of taxa. The resultant matrix is so strongly informed by this *a priori* system that it is indeed a "sanitized" distillation of all the traits that are possible to consider (Faith and Cranston, 1991). This distillation may connote subjectivity, but it also provides informed, often justifiable constraints on the analysis. Our knowledge of the marked variation in pelage color among species of mammals obviates the use of this character system in the higher-level phylogenetics of mammals. In contrast, our motivation in molecular studies is, as Patterson (1989, p. 473) notes, largely a matter of getting the data and processing them, because we are still in an early phase of exploration for genetic clues to phylogeny. Morphological data sets show structure by design; in molecular analysis, at our present state of knowledge, there is little choice in the matter. Thus it is not surprising that sequences of certain molecules (e.g., alpha hemoglobin for mammalian orders) yield phylogenetic patterns or trees that are even noisier (have more homoplasy) than patterns spawned by a random mixing of the same characters and taxa (Faith and Cranston, 1991). Just the same, we are at a stage where rational programs can be devised for sampling genes bearing on a particular problem. There is no excuse for ignoring the potential hierarchical structure of data whether morphological or molecular. Molecular evidence should not be blindly sampled without an effort to discriminate evidence that monitors low-level variation from evidence that reveals higher-level branching.

It has been claimed that homology statements in morphology are also enriched with independent (sometimes called direct) information, primarily ontogeny and paleontology. The status of these direct approaches for elucidating phylogeny is, however, a matter of debate. Ontogeny, though a powerful means of elucidating the similarity and transformation of characters, cannot be cut off (at least unambiguously) from assumptions involving, as in outgroup comparison, some framework of higher-level relationships (Eldredge and Novacek, 1985; Wheeler, 1990). The fossil record also has obvious shortcomings as a consistently reliable data source, although it does, in some instances, reveal a very compelling picture of history (see further comments below). Whatever the importance attached to ontogeny and paleontology, it should be noted that these lines of evidence are certainly not excluded from the molecular realm (e.g., Golenberg et al., 1990). Moreover, there are disclosures about molecular properties and dynamics (e.g., prevalence of certain transversions or transitions, secondary and tertiary structure etc.) that are of considerable value in applying sequence data to phylogenetic reconstruction (e.g., Wheeler and Honeycutt, 1988).

TREE CONSTRUCTION: PARSIMONY AND WEIGHTING

The basic methods for sorting out patterns of relationships by now comprise a menu of some familiarity. The list includes distance analyses (e.g., unweighted pair group methods, transformed distance, Fitch and Margoliash, distance Wagner, Neighbor joining, and others), parsimony (or maximum parsimony), evolutionary parsimony, and maximum likelihood (see Nei, 1991, for a recent review). Among these, maximum parsimony is distinctive for entailing only one assumption–that the least complex explanation of the data is the preferred one. Powerful computer programs such as PAUP (currently version 3.0, Swofford, 1991b) and HENNIG 86 (Farris, 1988) have catalyzed a widespread application of parsimony to complex data sets.

Parsimony may not yield the most accurate trees under certain assumptions concerning the nature of evolution (Felsenstein, 1978; Nei, 1991; Sidow and Wilson, 1991), but such qualities of evolution are not of course known *a priori*. Maximum likelihood carries with it the cost of extra assumptions. Given our ignorance, it seems advisable to minimize the assumptions necessary to carry out the reconstruction (see also Cracraft and Helm-Bychowski, 1991). Perhaps the only instance of some *a priori* justification thus far is the knowledge of transition/transversion bias in nucleotide sequences. Evolutionary parsimony (Lake, 1987) and compositional statistics (Sidow and Wilson, 1991), techniques sensitive to this bias, have been applied as an alternative to maximum parsimony. Their practical application is unfortunately limited to four or five taxa (Lake, 1988). Moreover, evolutionary parsimony assumes equal probabilities for certain transversions (e.g., probability of G to T = G to C) that are inappropriate for a significant range of gene sequences (Sidow and Wilson, 1991).

The application of parsimony is clearly susceptible to modification, which usually entails some form of character weighting. In molecular studies weighting is now employed as a means of accounting for varying properties of molecules, such as the tendency for more frequent substitutions in the form of transitions (see Miyamoto and Boyle, 1989; Cracraft and Helm-Bychowski, 1991). The application is well illustrated by analysis of molecular data bearing on the question of bat origins and relationships. Adkins and Honeycutt (1991) recently analyzed important sequence data for the mitochondrial cytochrome oxidase subunit II gene (COII) in representatives of both bat suborders (a megachiropteran and a microchiropteran), primates (*Galago* and *Homo*), tree shrews (*Tupaia*), and flying lemurs (*Cynocephalus*). These are taxa traditionally associated in varying sequence within the superorder Archonta (Gregory, 1910; Novacek, 1990). Also included were three mammalian taxa representing edentates (the armadillo *Dasypus*), rodents (*Mus*), and artiodactyls (the common cow, *Bos*), three orders not customarily lumped within Archonta. Adkins and Honeycutt (1991) focussed primarily on whether mega- and microchiropterans were either diphyletic (Smith and Madkour, 1980; Pettigrew, 1986) or monophyletic (Wible and Novacek, 1988; Baker et al., 1991; Simmons et al., 1991). They were also concerned with relationships of bats to other archontans and with the validity of the Archonta itself. They chose as the outgroup the armadillo *Dasypus*, an appropriate selection given the assertion that edentates are a remote branch relative to other clades of eutherian (placental) mammals (McKenna, 1975; Novacek, 1982). (It is, however, evident that this early split for edentates is ambiguous enough [Novacek, 1990] to consider the possibility of alternative outgroups for rooting the tree. The likely candidates in the COII data set are *Mus* and *Bos*. When these are alternated for *Dasypus*, only minor topological changes involving the relative positions of the outgroup candidates result with obviously no change in the length of the tree.)

Results of the COII analysis clearly show the effects of different weighting strategies. Maximum parsimony of all equally weighted substitutions revealed a tree (=phylogenetic pattern as used here and below) in which bats were not monophyletic (Fig. 1a). Given the expectations that mtDNA sequences show transitions (especially at the third cordon position) that approach saturation at remote divergence times (Brown et al., 1982), Adkins and Honeycutt (1991) also considered a parsimony analysis restricted to transversions only. The resultant tree supported bat monophyly (Fig. 1b). The authors viewed the topology produced by transversion parsimony as a better representation of the COII evidence.

Restriction to transversion substitutions in the COII sequences and other molecular data is founded on the high ratio of transitions over transversions observed in mitochondrial genes. Moreover, such high ratios apply to COII sequences in primates (see Adkins and Honeycutt, 1991). Nonetheless, some ambiguities concerning this weighting procedure remain. Should, for example, the transitions in COII sequences be excluded in the case of bats and their putative relatives? If not, what weights should be assigned to either transitions or transversions? One might assume that transitions of certain gene sequences will not retrieve very ancient splitting events, but the conclusions regarding one set of sequence data may not be readily transferable to another. A study of mitochondrial (ribosomal) genes by Miyamoto and Boyle (1989) in fact showed that transition/transversion ratios varied between 13.71 and 0.64 in pairwise comparisons for eight species of mammals. I found that weighting bias for the transversion/transition ratio of 10/1 or 5/1 maintained the all-or-none scheme shown in Figure 1b, but a weighting ratio of 2/1 produced a topology resembling the unweighted result (Fig. 1a) in promoting fission of the bat clade (Fig. 1c). It therefore seems advisable, unless *a priori* evidence otherwise dictates, to apply a spectrum of weights to sequence data and at least account for the threshold values that affect topologies.

Another means of weighting sequence data involves *a priori* combinatorial analysis of aligned sequences (Wheeler, 1990). Here weights are assigned simply as a function of the frequency of co-occurrence of nucleotides at variable positions. This derives from the argument that the greater the frequency that two nucleotides occur in the same position, the higher the probability of interchange (Wheeler,

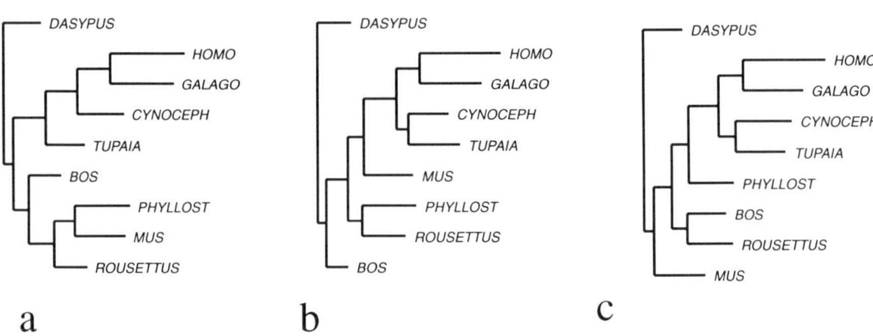

Figure 1. Most parsimonious trees for cytochrome oxidase II (COII) substitutions for a megachiropteran (*Rousettus*) a microchiropteran (*Phyllostomus*) and selected mammalian taxa. Solutions are based on: (a) all substitutions equally weighted; (b) transversions only; (c) transversion/transition weighting ratio of 2/1.

1990). An alternative weighting of nucleotide combinatorial transformations involves an interactive dynamic approach that draws on topology to assess weights (Williams and Fitch, 1989; Fitch and Ye, 1991). These approaches do not, however, decouple procedures from some problematic assumptions. Wheeler's (1990) combinatorial method, for example, assumes that no intermediate state changes are unrecorded in the data matrix. We can see, in the case of rapidly evolving gene sequences for a set of distantly related taxa, that such an assumption may be less than secure.

But if there is risk in applying weighting to a set of molecular sequence data, that risk seems certainly greater with morphological characters. Despite some arguments to the contrary (Szalay and Bock, 1991), I see little or no basis for predicting the evolutionary tendencies of a set of morphological traits independent of branching solutions where such characters are applied (also see Eldredge and Novacek, 1985). *A priori* weighting in morphology simply isn't yet credible. Alternatively one might apply *a posteriori* weighting that favors those characters with a minimum number of contradictory changes in the form of homoplasy. Successive weighting (Farris, 1988) has been widely adopted for this purpose. If the objective is to force a greater level of resolution or a more constrained set of solutions, successive weighting affords clear advantages. Figure 2 shows results of unweighted and weighted parsimony analysis of 88 morphological traits (see Novacek, 1992) in 20 extant mammal orders and four fossil taxa (three very poorly represented fossils in the original data set of Novacek, 1992, were excluded for heuristic purposes). Unweighted parsimony yielded 96 solutions with a consensus tree showing limited resolution (Fig. 2a). Successive weighting reduced the equally parsimonious solutions to 18 trees and provided greater resolution (Fig. 2b). Internal nodes increased from 16 to 20. Resolution, for example, includes a grouping of the extant archontan clades, and basal branching of the edentate-pholidote clade, a result also produced if only the extant orders are considered (Novacek, 1992).

Reliance on this kind of operation does, of course, assume that gains in stability and greater resolution are what we are after. Again

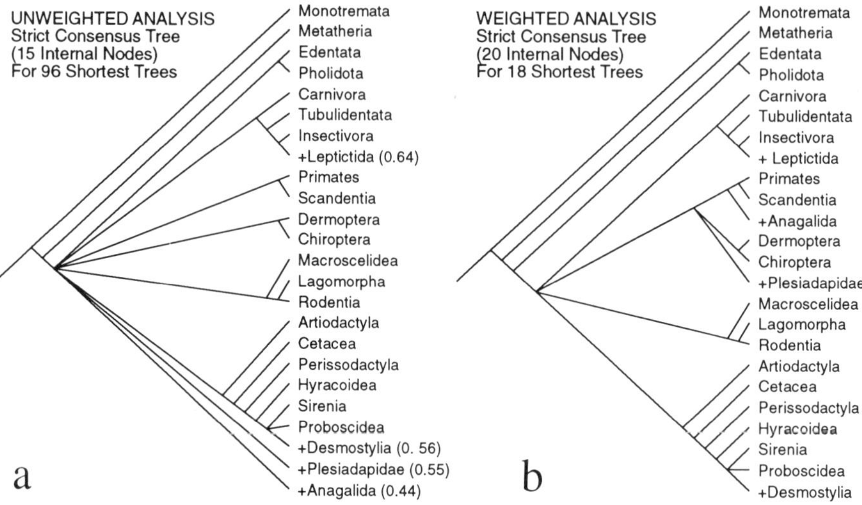

Figure 2. Strict consensus trees of most parsimonious cladograms for 20 extant and four extinct clades of mammals (for data, see Novacek, 1992). (a) Unweighted solution; (b) solution based on successive weighting. Numbers in parentheses indicate the proportion of total characters represented in the fossil.

there is no *a priori* basis for higher weights; merely a scheme that is sensitive to an inverse function of the homoplasy in the data known from the unweighted tree. Moreover, successive weighting with morphological data does entail some risks. Characters not uniformly represented (such as many characters in fossils or highly transformed taxa) may be artificially up-weighted. Consider again the analysis of higher mammalian clades shown in Figure 2. Maximum weights were assigned for numerous characters in which there was no representation or very poor representation in the fossil taxa. The majority of characters actually known in the fossils received lower weightings based on the degree of homoplasy they exhibited. Clearly, the potential homoplasy in the data matrix is reduced by the large subset of missing entries, and this imperfection is exploited by the successive weighting operation. Such an approach should be used judiciously, particularly in cases incorporating fossils or other data sets where the proportion of missing or ambiguous entries is large. It should be noted, finally, that some degree of *a priorism* in weighting a morphological data set, is unavoidable. This is because the characters analyzed represent those chosen from a universe of possibilities.

SAMPLING CHARACTERS: OPTIMAL GENES

Analyses like those described above can always be embellished by further sampling. Here, however, a dilemma arises. Given practical limitations, do we seek more information about characters or simply add additional taxa to the original character pool? The problem eludes a general prescription, but see Wheeler (1992). In molecular analysis this dilemma is particularly relevant because, despite the availability of PCR (polymerase chain reaction) techniques, comprehensive sampling is a formidable enterprise. Although there is an obvious need to sample more taxa, perhaps the most intriguing aspect of the molecular problem concerns the selection of the character set; namely a particular suite of genes or their protein products.

Data have accumulated sufficiently to muster a few generalizations about this sampling effort. Comparative sequencing has concentrated on ribosomal RNA genes in bacteria, (Woese, 1989; Lake, 1989), broad level studies of eukaryotes and metazoans (Raff et al., 1989; Sogin et al., 1989), plants (Eckenrode et al., 1985; Zimmer et al., 1989), and insects (Wheeler, 1989). Discrepancies are evident among different studies of similar gene sequences (e.g., bacteria) and between molecular and morphological results (higher eukaryotes, metazoans, flowering plants, insects), but some level of correspondence has been achieved. Wheeler (1989), for example, found that congruence between rDNA trees and morphologically based trees (e.g., Kristensen, 1981) was apparent if a combination of sequence changes and restriction site changes were applied to nodes on the molecular tree.

The intensity of sequencing effort in vertebrates roughly tracks a scala naturae from fish to human. Studies of nonmammalian branches largely concern rRNA genes (Hillis and Dixon, 1989; Mindell, 1991), whereas mammalian taxa, especially primates, have been lavished with attention based on sequences for a substantial number of mitochondrial and nuclear genes (Holmes, 1991). These have followed up on comprehensive efforts to sample proteins in diverse mammals. Amino acid sequences in several proteins (primarily lens alpha crystallin, alpha hemoglobin, beta hemoglobin, and myoglobin) reveal mixed results. Alpha lens crystallin produces a strong signal with a high consistency but the clearest supported nodes are limited to a sector of the mammalian tree (De Jong, 1982; McKenna, 1987, 1992; Wyss et al., 1987). The hemoglobins and myoglobin produce trees of relatively higher homoplasy and topologies in marked conflict with each other and with lens alpha crystallin results (Wyss et al., 1987). In the direction of gene sequencing, several targets, such as 12S and 16S rRNAs, an assortment of tRNAs, cytochrome oxidase subunit II (COII), cytochrome b, ND4, and ND5, as well as the nuclear e-globin and ubiquitin genes, are now known for a reasonably diverse set of taxa (Brown et al., 1982; Hayasaka et al., 1988; Miyamoto and Boyle, 1989; Ammerman and Hillis, 1990; Kraus and Miyamoto, 1991; Irwin et al., 1991; Adkins and Honeycutt, 1991; Vrana, in prep.).

The molecular database in mammals has underpinned some general assessments of DNA sequence data. Mitochondrial sequences have attracted the most attention, but these represent only 30 genes instead of minimally 50,000 genes in the mammalian nucleus (Gall, 1981). In addition, mitochondrial genes lack non-coding and other structural DNA's of potentially high phylogenetic value. More importantly, there are indications that mitochondrial genes evolve at rates high enough to saturate loci with changes over a comparatively brief span of evolutionary time (Brown et al., 1979). Hence, more ancient (e.g., interordinal) splitting events in mammals may not be retrieved through analysis of these sequences. Miyamoto and Boyle (1989) attributed such qualities to 12S and 16S rRNA and their adjacent tRNA genes. While these sequences were effective in yielding a branching pattern for artiodactyl taxa closely compatible with morphological results, saturation levels were too high for confident resolution pertaining to the extraordinal relationships of artiodactyls. Their plot of percent sequence divergence against times of divergence (derived from estimates of the fossil record) showed high saturation rates in the most recent 10 to 20 million years for total substitution and gap events and for transitions alone (Fig. 3). Transversions showed a more conservative profile, suggesting a greater potential for illuminating older divergences (Fig. 3). Likewise, Irwin et al. (1991) estimated explosive transition rates for the 10 to 20 million year span, especially

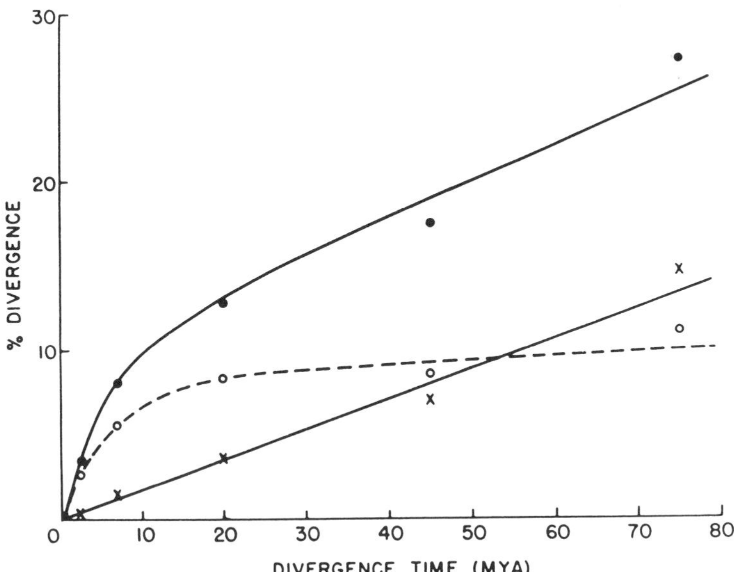

Figure 3. Percent sequence divergence against estimated times of divergence for rRNA and tRNA genes in selected artiodactyls and other mammals. Curves represent total substitutions and gap events (solid circles), transitions only (open circles), transversions only (x). From Miyamoto and Boyle (1989).

at third cordon position, in cytochrome b genes.

Elaboration of these comparisons is difficult given sampling limitations. From a sample of three taxa (primates, rodents, and artiodactyls), Holmes (1991) found that likelihood estimates based on an assumption of equal rates produced contradictory results between nuclear genes and mitochondrial genes. Moreover, Holmes (1991) suggested that rates of change for nuclear genome in rodents might be remarkably high. Thus, increased gene sampling leads one to a picture of mosaic evolution in different genes for different taxa that has its clear analogy in morphological work. Probing the genetic universe for optimal sequences is far from a trivial exercise.

Despite the difficulties posed in selecting the right genes in the right taxa, there is an obvious advantage to molecular sampling. The potential for tapping into a rich array of discrete traits in molecular work is enormous. The study noted above for COII deals with 684 different sites alone. One can therefore well imagine that molecular data has potentials for swamping out any signal produced by a much smaller set of morphological traits. Such effects, however, are minimized when weighting assumptions reduce the number or influence of informative molecular traits (see further discussion below).

SAMPLING TAXA: THE USE OF FOSSILS

The other pathway of improved sampling involves additions of taxa. Here we see considerable weight on the molecular side. Obviously, molecular work is taxonomically impoverished compared to studies of morphology. To draw much from this distinction is a mistake; vast improvement in the sampling for an increased diversity of genes in an expanded set of taxa is only a matter of time, money and effort. There is, however, one fairly intractable problem for the expansion of this molecular database: it cannot comprehensively draw on evidence from fossil organisms. Exceptions are, of course, dramatic retrievals of gene sequences in fossils that are either relatively recent or extraordinarily preserved (Golenberg et al., 1990; DeSalle et al., 1992).

This limitation is unfortunate because instances have been

revealed where fossils play a pivotal role in reorienting the phylogenetic relationships of living taxa. Perhaps best known in this regard is the case of higher amniote relationships, where the outcome is strongly affected by the rich morphological database represented by fossil synapsids (Gauthier et al., 1988). Trees based on data from extant taxa alone showed a grouping of crocodiles, birds, and mammals to the exclusion of turtles and lizards. Incorporation of a large number of fossil clades produced a sequence in strong conformity with the traditional pattern; mammals and their sister synapsid groups form a clade isolated from a reptile clade, and successive branching of turtles, lizards, crocodiles, and birds is interspersed with a variety of extinct clades (e.g., dinosaur subgroups). On the molecular side, Hillis and Dixon (1989) found that conserved sequences from 28S rDNA data in five extant taxa relevant to the amniote problem had limited power to produce a definitive result. All of the most parsimonious unweighted cladograms and two of the three most parsimonious weighted cladograms place birds (namely *Cacatua*) with mammals (*Mus*) to the exclusion of squamate reptiles (the lizard *Rhineura*). But trees conforming to the traditional arrangement were only two steps longer in the unweighted analysis and a single step longer in the weighted analysis. Hillis and Dixon suggested future improvements may come with sampling less conserved regions, closer outgroups and additional taxa, but acknowledged that DNA sequences in any extant groups may ultimately fail in retrieving some of the remote splitting events represented by fossils.

Do fossils indeed consistently increase chances of resolving ancient phylogenetic events? Since any added taxon, whether fossil or extant, can alter an original tree topology, a unique status for fossil data must hinge on other properties. The most common assertion in this vein is simply the alleged correspondence between age and primitiveness. Older fossils are more likely to preserve primitive conditions than are younger fossils or extant taxa (Simpson, 1975; Donoghue et al., 1989). Hence, older fossils are likely to illuminate ancient splitting events obscured in the mosaic of highly derived traits present in living taxa. Despite the intuitive appeal of this argument, cases do not in general mirror the dramatic effects of the amniote example (Donoghue et al., 1989; Novacek, 1992). There is, however, a means of inspecting, at least against independent cladistic results, the premise that ancientness and primitiveness are generally correlated. An examination of 24 cases derived from cladograms of vertebrate groups with known fossil records showed appreciable variation in the strength of this correlation (Norell and Novacek, 1992). There was, in fact, no indication that fossil age data will always provide for a reliable prediction of the degree of primitiveness. Yet correlations between the clade ranks (from higher level to successively lower level rankings) and dates of earliest occurrence for the taxa were notably high in a majority of cases examined. The potentials of fossil evidence to retrieve successive phylogenetic events seem notable, especially in cases (e.g., horses) where age and morphological data are abundant (Norell and Novacek, 1992).

But the formidable attributes of fossils in this context must, however, be linked to a truism about their limitations. Fossils are incompletely preserved, sometimes grossly so. This has been such a liability that it was at one time difficult to apply the same kinds of analytical methods to fossil data as one might readily use on extant taxa. Popular computer programs like PAUP and HENNIG 86 do, however, incorporate algorithms that account for missing data. Unfortunately the algorithms do not eradicate the problem of missing data; they merely provide a means of scoring the missing entries in a way compatible with the most parsimonious arrangement for the known characters. As noted above, this allows insidious bias on operations like successive weighting. Even unweighted analyses are affected, because there is always a likely bias against homoplasy in scoring missing entries to conform in the least problematic fashion with the available data (Platnick et al., 1991).

Another disadvantage of incorporating fossils or extant taxa with an abundance of missing entries is that the end results may hardly provide coherent patterns. Addition of seven fossil clades to a mammal data set comprising 20 extant orders largely diminish resolution and proliferate most parsimonious trees (Novacek, 1992). Successive addition of each of these fossil clades to the data set (Fig. 4) shows that the number of most parsimonious solutions increased dramatically with addition of the fifth taxon (similar trajectories are produced regardless of the order in which the fossil clades are added).

Despite these problems related to incomplete preservation, fossils

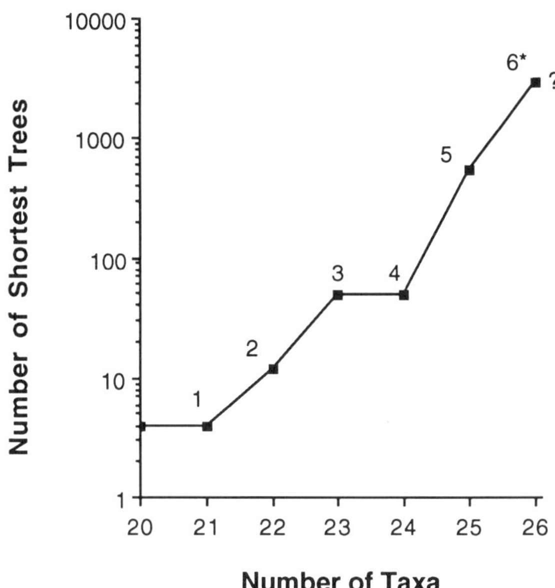

Figure 4. Proliferation of most parsimonious trees with successive addition of incompletely preserved fossil taxa. Trees are generated from an 88-character matrix for 20 extant mammalian orders (see Novacek, 1992). Numbers represent the following added fossil clades: (1) Leptictida; (2) Desmostylia + 1; (3) Plesiadapidae +2; (4) Microsyopidae + 3; (5) Kennalestida + (4); (6) Asioryctidae + 5. A seventh fossil taxon (Anagalidae) from the original data is not added here. Values for 1 through 5 are based on shortest trees found with the exact branch and bound algorithm in (PAUP). 6* is a heuristic solution with a default upper limit of 3000 trees. Actual number of most parsimonious trees could greatly exceed this value. Values for numbers of shortest trees vary depending on the order of the fossil taxa added, but marked proliferation occurs with a fifth taxon regardless of the addition sequence.

do provide illumination in certain sectors of the mammalian tree. For example, new disclosures on the skeletal features of an extinct alleged primate group, the Paromomyidae, suggest striking resemblance to specializations in living flying lemurs (Beard, 1990). This naturally inspires a claim that new fossil evidence favors close primate-dermopteran affinities (Beard, 1990; Adkins and Honeycutt, 1991; Novacek, 1992; Bailey et al., 1992) rather than dermopteran-bat affinities (Novacek, 1982, 1986; Thewissen and Babcock, 1991). Closer examination of the relevant morphological evidence supports the primate-dermopteran connection. Tables 1 and 2 display 49 characters derived from various morphological studies involving bats, primates, tree shrews, dermopterans, and fossil paromomyids (see Table 1 for explanation). For reference comparisons with molecular results (Fig. 1 and discussion below), the ancestral states for these characters in rodents, edentates, and artiodactyls are also provided. Unweighted parsimony analysis of the extant taxa produced two trees of high consistency (CI=0.794, RI=0.821) linking dermopterans to a monophyletic Chiroptera (Fig. 5a, b).

A test of these results incorporates new comparative data gathered by Beard (1993) for his study of paromomyids and other fossil archontans. Beard (1993) examined 29 morphological characters, three of which overlap with characters listed in Tables 1 and 2 herein. Combination of these two data sets (eliminating any redundancy in characters) can be applied to a reduced number of terminal extant taxa considered by Beard. I ran a branch-and-bound analysis of the combined data, including Paromomyidae, one of the three key fossil taxa considered by Beard (the others were the Micromomyidae and Plesiadapidae). The one shortest tree resulting from this expanded data set (Fig. 5c, tree length = 114 steps, CI = 0.804, RI = 0.766) showed a close sister grouping of dermopteran-paromomyid clade with primates to the exclusion of the bats and scandentians. In this case, the missing data for Paromomyidae did not prevent a clearly resolved solution. This solution contradicts the branching pattern (Fig. 5a, b) based on data (Table 1) not informed by characters covered in Beard's (1993) study of fossil archontans.

Table 1. Derived characters reviewed herein for support of phylogenetic relationships shown in Fig. 5. Sources of characters are Wible and Novacek (1988), Pettigrew (1986, 1991), Beard (1990), and references cited therein.

---

1. Pendulous penis suspended by a reduced sheath between genital pouch and abdomen [1]
2. Sustentacular facet of the astragalus in distinct medial contact with distal astragalar facets [2]
3. Complete postorbital bar
4. Petrosal-derived osseous canals around intratympanic portions of facial nerve and stapedial artery
5. Anterior carotid foramen in basisphenoid converted to a long tube
6. Tegmen tympani expanded anterolaterally to roof epitympanic recess; tegmen tympani with an epitympanic crest within which stapedial artery runs
7. Fenestra cochleae (round window) faces directly posteriorly
8. Subarcuate fossa greatly expanded and dorsal semicircular canal clearly separated from the endocranial wall of the squamosal
9. Neural spines on cervical vertebrae 3-7 weak or absent [3]
10. Ribs flattened, especially near their vertebral ends [3]
11. Forelimbs markedly elongated [3]
12. Patagium continuously attached between the digits of the manus
13. Ulna markedly reduced at both its proximal and distal ends [3]
14. Humeropatagialis muscle inserts into plagiopatagium
15. Premaxilla greatly reduced
16. Jugal greatly reduced
17. Tegmen tympani tapers to a slender process that projects ventrally into the middle-ear cavity medial to the epitympanic recess
18. Ramus inferior of the stapedial artery passes through the cranial cavity dorsal to the tegmen tympani
19. Ramus infraorbitalis of the stapedial artery passes through the cranial cavity dorsal to the alisphenoid
20. Two entotympanic elements in the floor of the middle ear cavity: a large caudal element and a small rostral element grooved by (or forming a canal around) the internal carotid artery
21. Distal end of humerus lacks supratrochlear depression and supinator ridge [3]
22. Entepicondylar foramen of the humerus absent [3]
23. Digits 2-5 of manus greatly elongated to support wing membrane [3]
24. Manus rotated 90° from position typical of quadrupedal mammals
25. Claws lost from terminal phalanges on digits 3 and 4 of forelimb [3]
26. Hindlimbs rotated from typical quadrupedal orientation (femur directed laterad and tibia directed caudad during flight and dorsolaterad and ventrad, respectively, during crawl)
27. Trochlea of astragalus lacks medial and lateral marginal ridges [3]
28. Calcanealastragalar facet of calcaneum modified from convex process to depression or trough [3]
29. Peroneal tubercle of calcaneum absent [3]

30. Depressor osseous styliformes muscle runs between calcaneum and calcar
31. Occipitopollicalis muscle present on the leading edge of propatagium
32. Sartorius muscle absent
33. Preplacenta broad and horseshoe-shaped with definitive placenta more localized
34. Definitive yolk sac "gland-like"
35. Prominent "interstitial membrane" in the chorioallantoic placenta
36. Cortical somatosensory representation of forelimb reverse of that in other mammals
37. Accessory cavernosus tissue absent
38. Enlarged neocortex
39. Reduced, but binocularly balanced, retinotectal pathway subserving only contralateral hemifield of visual space [4]
40. Ipsilateral retinal input to rostralmost tectum [4]
41. Ipsilateral bias in retinotectal-hypothalamic pathway [4]
42. Reduced accessory optic input to medial terminal nucleus [4]
43. Complex of characters in organization of lateral geniculate nucleus, with segregation of ocular input to different laminae and with magnocellular laminae adjacent to optic tract
44. Distally expanded, vascular corpus spongiosum
45. Derived, noninsectivorous dentition
46. Distal radius and lunate expanded
47. Complete petrosal bulla
48. Rostral wing muscles innervated by facial nerve [5]
49. Rostral wing muscles with double innervation by facial and cervical spinal nerves [5]
50. Elongate intermediate phalanges
51. Lunate position distal to scaphoid
52. Secondary synovial joint between sustentaculum tali and medial astragalar peli

---

[1] Although it is doubtful that abdominal testes are primitive for Microchiroptera, some members of that group have this trait. Hence, character 1 could be scored as polymorphic for Microchiroptera, but with no change in topology (see Table 2 and Fig. 5).
[2] The drastic alteration of the microchiropteran astragalus involves presumed loss of the sustentacular facet. Precursor state for character 2 is unknown in Microchiroptera.
[3] Although a variety of skeletal features might be correlated with development of the wing (character 12), these skeletal traits need not be exclusive to volant mammals. Conversely, many of these traits are not found in other gliding mammals (see Baker et al., 1991).
[4] In recognition of the bat diphyly argument characters 39-43 are analyzed separately. As with wing features, however, some of these might be combined as integrated components of the retinotectal pathway. Note that based on Thiele et al. (1991), many of these features are polymorphic for Megachiroptera (Table 2).
[5] The gliding squirrel *Glaucomys* also shows innervation of the patagial muscles by the facial nerve. However, it is highly unlikely that the glissant structures and the concomitant facial nerve modifications are primitive for Rodentia. Characters 48 and 49 are scored as primitively absent in rodents (Table 2). Polymorphic coding of characters 48 and 49 does not alter the results discussed here.

Table 2. Primitive (0) and derived (1) states for 52 morphological characters in the bat suborders and selected mammal groups. ? = missing or ambiguous entry. P = state coded as polymorphic. Characters are described in Table 1.

TAXA	CHARACTERS [1] 0000000001111111111222222222233333333334444444444555 1234567890123456789012345678901234567890123456789012
Edentata	00000000000000000000000000000000000000000000?0000000
Anthropoid	1111110000000100000000000000000000000111111111100000
Strepsirhine	1111110000000100000000000000000000000111111110100000
Dermoptera	1100001111111000000000000000000000000111111111010111
Scandentia	1111110000000000000000000000000000000110000000000000
Artiodactyla	000000000000000000000000000000000000?000000?10000000
Microchiroptera	1?00001111111111111111111111111111111100000000011000
Megachiroptera	11000011111111111111111111111111111111PPPPP111011000
Rodentia	0000000000000000000000000000000000000000000000000000
Paromomyidae	?10???00??11??????????00000000??????????????1????111

[1] Polymorphic coding (P) for character 1, Microchiroptera and characters 48, 49, Rodentia (see Table 1), yield topologies identical to that in Fig. 5c, d, but trees have slightly greater length (76) and slightly lower CI (0.789) and RC (0.637)

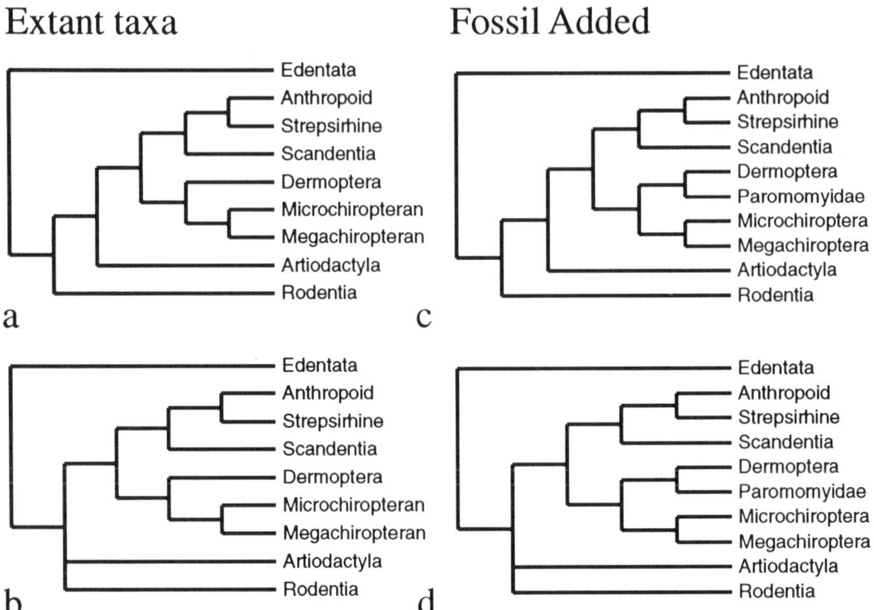

Figure 5 (a, b). Most parsimonious trees for extant taxa listed in Table 2. Solutions based on characters 1-49 (in Table 1). Tree lengths = 68; CI (consistency index) = 0.794; RI (retention index) = 0.821; RC (rescaled consistency index) = 0.652. (c). Most parsimonious tree including fossil taxon Paromomyidae for data in Table 1 combined with 26 characters from Beard (1993). Three of Beard's 29 characters are already listed in Table 1 (characters 3, 4, and 6). Terminal taxon set reduced to conform with Beard's coverage. For description of additional characters, see Beard (1993). Tree length = 114 steps, CI = 0.804, RI = 0.766, RC = 0.628.

With respect to the problems of missing data, discussion should not be confined to fossils. Data for extant taxa can, in effect, be equivalent to missing if a character is so transformed that it must be scored as ambiguous (Gauthier et al., 1988). Living mammals show such marked transformations in certain features of the basicranium, auditory region, and jaw joint, that they provide little or no evidence that serves to discriminate the synapsid-mammal clade from other amniotes (Gauthier et al., 1988). Missing data in molecular work is largely a matter of uneven sampling for taxa. Miyamoto and Goodman (1986) used tandem alignments to combine the evidence of several proteins not uniformly represented in the large number of mammalian species compared. Although this approach had the advantages of allowing such a taxonomically diverse treatment, it admitted a data matrix with an appreciable level (33%) of missing information. This factor, coupled with the inconsistencies among different protein data (Wyss et al., 1987), might be expected to diminish the phylogenetic signal produced by the combined protein data. This is also apparent when the protein and morphological data are combined. The shortest tree favored by tandem alignment parsimony of the protein data showed significant discrepancies with morphologically based trees (Miyamoto and Goodman, 1986; Novacek et al., 1988). But when I combined data from the four best sampled proteins (81 characters binary coded for changes following Wyss et al., 1987) with the anatomical evidence (88 characters), much of the structure favored by the morphological tree was preserved (Fig. 6). While morphological data here, on face value, show more internal consistency, the lack of information in a large part of the protein data certainly diminished the impact of the latter.

COMPARING RESULTS: CONGRUENCE, CONFLICTS AND CONSTRAINTS

The recognition of only one possible history for organisms raises the expectation that any kind of information–whether from genes or fossils–will converge on the pattern that discloses this history. Congruence among different results is therefore a working principle in phylogenetic analysis. The degree of congruence also suggests something about the relative strength of the phylogenetic signal. In the bat case, some morphological evidence (Wible and Novacek, 1988; but see Pettigrew, 1991), immunological data (Cronin and Sarich, 1980), independent studies of amino acid sequences for three globin genes (Czelusniak et al., 1990) sequences in COII genes (Adkins and Honeycutt, 1991), e-globin (Bailey et al., 1992), 12S rRNA (Ammerman

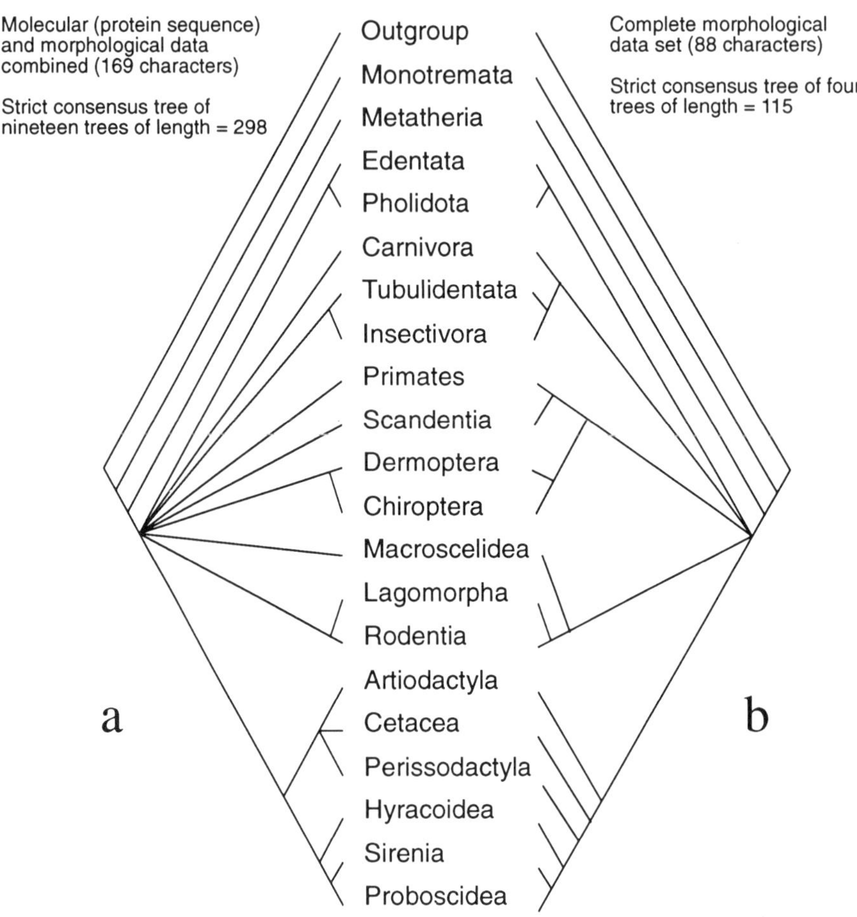

Figure 6. Strict consensus trees for combined morphological and molecular (protein sequence) data (a) and morphological data only (b) for 20 orders of extant mammals. Morphological data from Novacek (1992), binary coded protein sequence data from Wyss et al. (1987).

and Hillis, 1990; Mindell et al., 1991) and COI (Mindell et al., 1991) all point to monophyletic origin of the living bat suborders and a single origin for powered flight in mammals. (It should be noted that the assumptions underlying these correspondent results do, in some cases, differ. The favored COII result is based on transversion parsimony, while the e-globin result is based on [presumably] an unweighted analysis of all substitutions. A sampling of results for the latter based on different weighting schemes would be illuminating.) A weaker signal, however, must currently apply to the discovery of the nearest neighbors to bats. The morphological support for a close bat-flying lemur relationship (Wible and Novacek, 1988) is not matched by either immunological data (Cronin and Sarich, 1980) or gene sequence data (Adkins and Honeycutt, 1991; Bailey et al., 1992). Conflicts over the relationships of primates, dermopterans, and tree shrews also apply to the different molecular results, although there is general agreement from the molecular side that bats are a remote clade relative to other archontan taxa (Fig. 7).

But conflicts among hard-won results are not necessarily a cause for despair. Such conflicts do offer insights on the problems underlying the competing approaches. One way to investigate these inconsistencies is to constrain an analysis–whether molecular or morphological–where independent evidence might suggest such constraints. Some morphological and molecular data, for example, clash with respect to the recognition of the superordinal grouping Archonta (Gregory, 1910) to include primates, bats, scandentians, and flying lemurs. Both the unweighted and weighted analyses of COII genes (Adkins and Honeycutt, 1991) did not maintain an archontan grouping. Although the morphological evidence for archontan monophyly is not particularly enriched (Novacek, 1990), the close connection between rodents and a primate-flying lemur-tree shrew clade to the exclusion of bats in the CO II result (Fig. 1b) is an unexpected contradiction. In fact, the weighted molecular result is consistent with morphological characters supporting archontans that can be applied only more ambiguously to include bats. However, the COII evidence relevant to this question seems potentially ambiguous. I constrained the topology to maintain monophyly for the

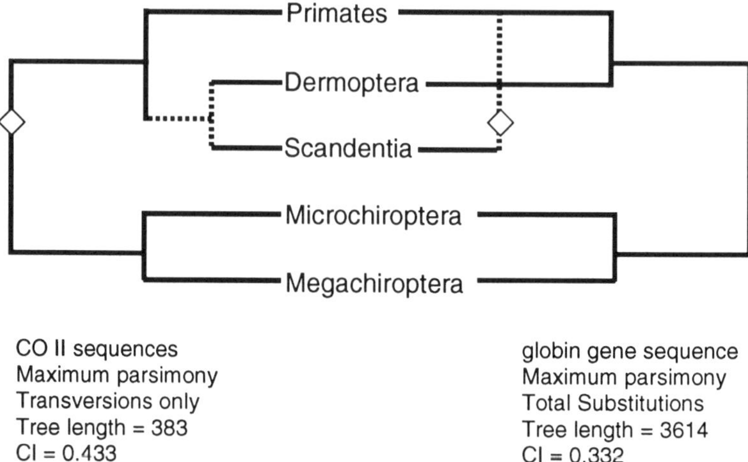

Figure 7. Most parsimonious topologies for COII sequences (Adkins and Honeycutt, 1991) and globin gene sequences (Bailey et al., 1992) for the bat suborders and putative relatives. Dashed lines indicate conflicting branching relationships for Scandentia (tree shrews). Diamond indicates a junction with a taxon not common to both data sets.

archontan groups (without further constraints) and found a weighted transversion result (Fig. 8a) only one step longer than the unconstrained tree (Fig. 1b). The constrained tree in the global parsimony analysis was only six steps longer; not an appreciable difference given a sampling of over 230 potentially informative sites.

The morphological data set (Tables 1 and 2) for the clades considered in the COII study produced a stronger signal for the preferred topology. The data set incorporates both characters supporting and characters rejecting bat monophyly, although some of the latter must be scored as polymorphic for the terminal taxa based on recently published studies (Thiele et al., 1991; see comments, Table 1). In this data set, edentates, rodents and artiodactyls show little or no evidence that would suggest their special affinities with one or another archontan clade. The hierarchy of character distribution is well

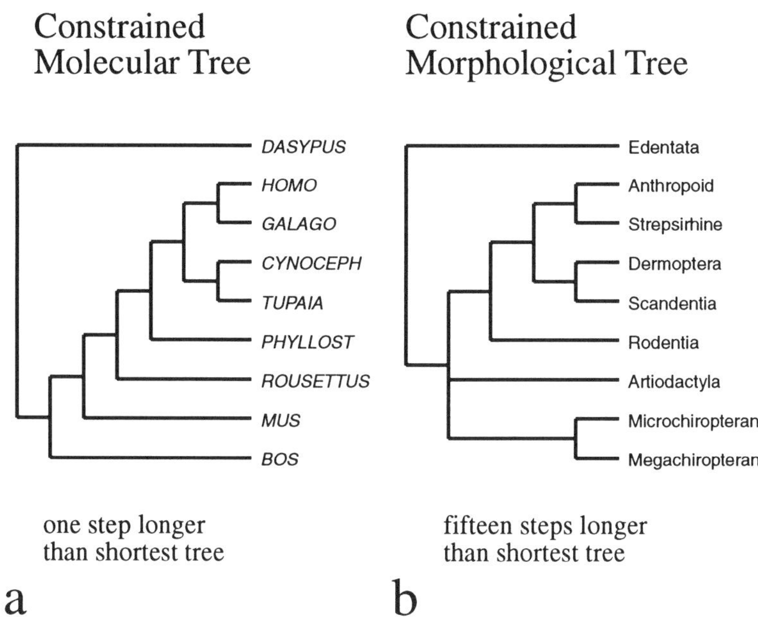

Figure 8. Constrained tree solutions. (a) Most parsimonious tree for COII transversions only (Adkins and Honeycutt, 1991) constrained to preserve the monophyly of Archonta. Tree is only one step longer than the unconstrained solution (Fig. 1b). (b) One of two most parsimonious trees based on morphological data (Table 1, characters 1-49) constrained to preserve the branching sequence in the transversion parsimony molecular tree (Fig. 1b). Tree is 15 steps longer than the unconstrained morphological trees (Fig. 5a, b). The equally parsimonious alternative to Fig. 8b places Artiodactyla outside Rodentia and the archontan clades.

established for the groups in question, even though Archonta are not supported by a large number of morphological traits. This is borne out by parsimony analysis of the data, where the shortest trees show a high consistency index of 0.794 (Fig. 5a, b). If the topology of the tree is constrained to conform with the weighted parsimony result of the COII study, the required tree (Fig. 8b) is 15 steps longer than the unconstrained results (Fig. 5a, b). A tree simply constrained for the insertion of rodents between bats and other alleged archontans is at least nine steps longer than the unconstrained result. Given a data set of only 49 potentially informative characters, one might judge the tree supporting archontan monophyly to be much more consistent with the morphology sampled than those trees designed to conform with COII results. The converse is not true, as the trees produced by weighted transversion analysis of COII hardly differed in length whether or not they were constrained by morphological results (cf. Figs. 1a and 8a). Interestingly, the molecular tree constrained for archontan monophyly (Fig. 8a) shows representatives of the bat suborders as non-monophyletic. Here it seems feasible in evaluating the molecular results to consider the structure suggested by other evidence.

EVALUATING RESULTS

Evaluating different data goes beyond simply comparing their resulting topologies. Proposed evaluations include inspection of consistency indices and other measures of the fit between the branching sequence and the original data set, bootstrapping and jackknifing, statistical significance tests, testing against randomized data (e.g., the permutation tail probability or PTP test), and evaluations of tree length distributions (all tree histograms). I shall not attempt to elaborate on several recent discussions of these methods (see Faith and Cranston, 1991; Cracraft and Helm-Bychowski, 1991; Li and Gouy, 1991; Hillis, 1991). Randomization tests like PTP (Faith and Cranston, 1991) have proven useful in discriminating problematic and ambiguous data. For example, alpha hemoglobin sequences showed a surprising inability to produce trees that were shorter than those derived from a random mixing of characters and taxa, whereas morphological data, lens alpha crystallin and other proteins showed significant and comforting departures from randomness (Faith and Cranston, 1991). Most results surveyed by this technique, however, seem to perform well against a test that places minimum expectation on the data (i.e., that a significant number of shorter trees than the "actual" shortest tree will not be found). In a sense, tree length distributions (Hillis, 1991) represent finer-grained analyses; they evaluate the strengths of particular groupings not only on the basis of the shortest tree but on the stability of these groups in trees of greater length (obviously there is some overlap here with PTP and other randomization tests, as distributions for actual trees can be compared against tree length distributions from randomized data).

The histogram approach is a refinement of various studies (e.g., Miyamoto and Boyle, 1989; Bailey et al., 1992) which evaluate the strengths of groupings on the basis of how many steps will promote group collapse. Hence Bailey et al. (1992) emphasized that the bat suborders in the shortest e-globin tree were tightly knotted, as 31 additional steps (nucleotide substitutions) were required to break up this grouping. When tree length distributions (histograms) are evaluated a statistic g_1 can be applied which measures the skewness of tree-length distributions for n trees of length T

$$g_1 = \sum_{i=1}^{n} (T_i - \bar{T})^3 / n s^3$$

where s is the standard deviation of tree lengths. This statistic in combination with the histogram profile can be useful in assessing the strength of phylogenetic signal (Hillis, 1991). For comparison with molecular results, I restricted the morphological data set for bats and their relatives (Tables 1 and 2) to the alleged archontan taxa and the edentate outgroup. The single shortest tree expectedly preserves chiropteran monophyly and dermopteran-chiropteran affinities (cf. Fig. 5). A histogram for all trees shows a distribution with a marked left hand skewness and a high negative g_1 statistic (Fig. 9a), properties that suggest a strong signal wherein one or a few trees are "better" than the rest (see Hillis, 1991). This is borne out by examination of group stability with increasing tree lengths; it takes seven steps to collapse Volitantia (the traditional superordinal category for dermoptera + bats)

Frequency distribution of tree lengths: Morphology

```
/-------------------------------------------
67 |# (1)
68 | (0)
69 | (0)
70 |## (2)
71 |## (2)
72 | (0)
73 | (0)
74 |##### (5)
75 |# (1)                 Collapse of Volitantia (Dermoptera plus Bats)
76 |### (3)
77 |## (2)
78 |########## (10)
79 |#### (4)
80 |##### (5)
81 |######## (8)
82 |######## (8)
83 |##### (5)
84 |##### (5)
85 |######### (9)
86 |################# (17)
87 |#### (4)
88 |# (1)                 Collapse of Chiroptera
89 |### (3)
90 |############ (12)
91 |# (1)
92 |####### (7)
93 |##### (5)
94 |## (2)
95 |### (3)
96 |######### (9)
97 |#### (4)
98 |#### (4)
99 |################## (18)
100 |###################### (22)
101 |################ (16)
102 |###### (6)
103 |############ (12)
104 |################## (18)
105 |############################################ (44)
106 |############################## (30)
107 |#################### (20)
108 |############################################# (45)
109 |############################################################ (60)
110 |############################# (29)
111 |##################### (21)
112 |############################################ (44)
113 |#################################################################### (68)
114 |############################################################## (62)
115 |######################################## (40)
116 |######################################### (41)
117 |############################################################ (60)
118 |################# (17)
119 |#### (4)
120 |################################## (34)
121 |##################################################################### (69)
122 |####################### (23)
\-------------------------------------------
```

g1= -1.268272

Figure 9a. Histograms of tree lengths based on exhaustive searches. (a) Distribution for morphological data (Tables 1, 2, Characters 1-49) representing seven taxa (archontan clades + Edentata).

and twenty steps to collapse Chiroptera (Fig. 9a). These are significant values given the small size of the original character set. An exact enumeration of all unweighted parsimony trees for COII substitutions and for the same representative groups showed appreciably less skewness (Fig. 9b) and a very low negative g_1 value. Here the bat suborders are non-monophyletic in the shortest tree, paraphyletic in the tree one step longer, and monophyletic in two trees two steps longer. The weak signal suggested by this distribution substantiates doubts expressed by Adkins and Honeycutt concerning the value of the unweighted COII substitutions data. Tree distributions kindly provided by Ron Adkins (pers. commun.) for the original COII data for nine taxa and the morphological data listed herein (Table 1, paromomyids excluded) show similar results. Both the unweighted and transversion only COII data produce histograms of low negative g_1 (-0.2374 and 0.2371 respectively), while the morphological data in the histogram for the 9 taxa produces a high negative g_1 (-1.5845).

Tree length histograms as evaluation tools do have ambiguities. One might argue that comparisons between morphological and molecular data in this manner may obscure the fact that a highly skewed distribution is predetermined by the inherent structure of morphological sampling (see remarks above). Nonetheless, Hillis (1991) documented several molecular results showing a high negative g_1. Another issue concerns the effect of taxonomic enrichment on tree distribution. Taxa that introduce contradictory characters may weaken the basis for the original groupings even if they conform to the original topology of the shortest tree. Such taxa, just the same, may contribute important new information on character transformation (Novacek, 1992). Hence, the detrimental aspect of increasing homoplasy may be offset by the closer correspondence between the character evidence (changes or transformation events) and actual taxa. One can even envision trees that approach a condition where every change is actually represented by a distinct taxon. This would apply to cladograms involving a series of fossil stem taxa, such as cladograms for horses (MacFadden, 1992; Norell and Novacek, 1992). As taxa are added to a data set, consistency diminishes and homoplasy tends to increase (Sanderson and Donoghue,

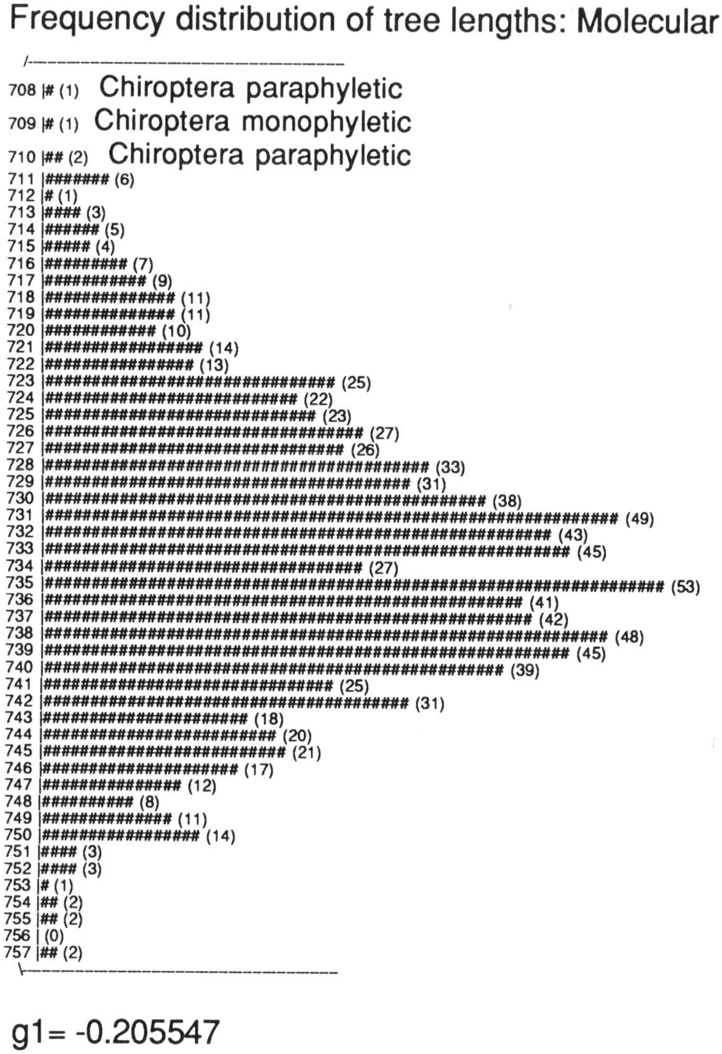

Frequency distribution of tree lengths: Molecular

```
 /-------------------------------------
708 |# (1)  Chiroptera paraphyletic
709 |# (1)  Chiroptera monophyletic
710 |## (2)  Chiroptera paraphyletic
711 |####### (6)
712 |# (1)
713 |#### (3)
714 |###### (5)
715 |##### (4)
716 |######### (7)
717 |############ (9)
718 |############### (11)
719 |############### (11)
720 |############# (10)
721 |################## (14)
722 |################# (13)
723 |################################# (25)
724 |############################# (22)
725 |############################### (23)
726 |#################################### (27)
727 |################################## (26)
728 |############################################ (33)
729 |########################################## (31)
730 |################################################### (38)
731 |################################################################## (49)
732 |########################################################## (43)
733 |############################################################# (45)
734 |#################################### (27)
735 |####################################################################### (53)
736 |####################################################### (41)
737 |######################################################## (42)
738 |################################################################# (48)
739 |############################################################# (45)
740 |##################################################### (39)
741 |################################# (25)
742 |########################################## (31)
743 |######################## (18)
744 |########################### (20)
745 |############################ (21)
746 |###################### (17)
747 |################ (12)
748 |########### (8)
749 |############### (11)
750 |################## (14)
751 |#### (3)
752 |#### (3)
753 |# (1)
754 |## (2)
755 |## (2)
756 | (0)
757 |## (2)
 \-------------------------------------
```

g1= -0.205547

Figure 9b. Distribution for molecular data (unweighted COII substitutions) representing seven taxa (archontan genera + *Dasypus*).

1989). One might therefore expect that histograms provide rigorous tests for taxonomically diverse data sets (although exact enumeration of tree lengths for more than a dozen taxa are in general computationally prohibitive–see Hillis, 1991). Nevertheless, shortest trees that account for a diversity of taxa have robust qualities that should be recognized.

COMBINING RESULTS: "THE TOTAL INFORMATION" APPROACH

As noted by Swofford (1991a), one of the most controversial issues in phylogenetics is whether or not data from diverse sources should be combined in a single data set. This approach is emphatically favored by Kluge (1989) who advocates the incorporation of "total information" or "total evidence" as the valid source of phylogenies and classifications. Swofford (1991a) expressed concern that the independent contribution of different data sets will be obscured if they are combined. Of course, neither operation precludes the other. Very often, properties of a given data set will even be illuminated when combined with another for purpose of analysis. For example, the insidious effects of missing data in the protein sequence set are all the more evident when combined with the more evenly represented morphological data set for extant mammalian orders (Fig. 6 and discussion above). Beyond reservations about losing information by combining data, there are concerns that the huge number of characters resident in nucleotide sequences will simply swamp out the signal from a smaller set of albeit compelling morphological traits.

Two factors, however, work against such an outcome. First, large sets of trait data (such as sequences not subject to weighted analysis) may have so much homoplasy that the outcome of the combined data hardly favors either this evidence or any other smaller and more informative data set added to the analysis. Second, weighting assumptions applied to large and complex data may reduce their impact in combination with smaller and highly informative data sets. Unweighted parsimony of all substitutions for the COII genes for bats and their candidate relatives draws on information from 684 sites (Adkins and Honeycutt, 1991). Combined with the 49 character morphological data set, the two shortest trees (956 steps; CI excluding uninformative characters = 0.557) show the influence of the morphological data set in maintaining a monophyletic bat grouping, but otherwise do not shift the molecular based topology (Fig. 1a) toward that supported by morphology alone (Fig. 5a, b). On the other hand, the transversion parsimony data set (that favored by Adkins and Honeycutt, 1991) clearly shows the influence of the morphological data in the combined analysis (Fig. 10). Here, the topology based on the combined data set is identical to one of the two shortest trees indicated by morphology except that the positions of representative Rodentia and Artiodactyla are interchanged (Fig. 5a). Thus, the integrity of dermopteran plus bat grouping supported by these morphological data is maintained even in the face of some contradictory molecular evidence.

CONCLUDING REMARKS

On the eve of a new age of comparative DNA sequencing, a volume edited by Colin Patterson (1987) was given the title "Molecules and morphology in evolution: conflict or compromise?" As we stand a few years down the road with a slug of more data in hand, how might we answer this question? When sampling of taxa and different genes for sequence data was poor, there was a natural resistance to the proclaimed value of such data in phylogenetic reconstruction. But it is now clear that such studies have and will continue to have profound effects on the course of systematics. In moving forward, it seems best to recognize morphology and molecular biology as providing compatible but distinctive databases. It is currently fashionable to criticize false dichotomies that pit molecules against morphology, fossils against recent data, etc. There are, however, essential differences that must be realistically acknowledged. Molecular data offer an enormous repository of characters but also present severe challenges to the search for informative characters. Morphological data, though less potentially diverse, offer a good deal of structure stemming from a sophisticated level of understanding of homology in diverse organisms. Morphological

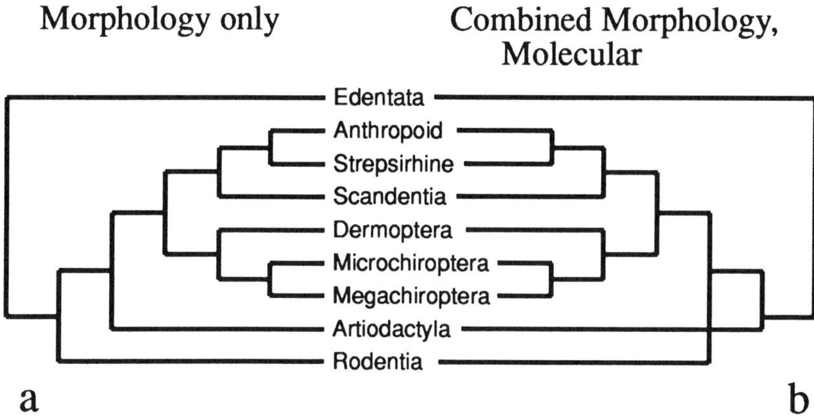

Figure 10. Morphology-based most parsimonious tree (a) versus combined morphological-molecular (CO II transversions only) tree (b). Topology of combined tree is identical to one of two shortest morphology based trees (Fig. 5a), except that positions of Rodentia and Artiodactyla are interchanged.

data are also uniquely effective in representing a large body of fossil evidence; an attribute that nonetheless comes with the problem of missing data due to incomplete preservation. Perhaps one of the great events in biology is that the learning curve for the analytical assessment for both kinds of evidence greatly accelerated in the last two decades of the 20th century. We shouldn't diminish our enthusiasm now for pursuing either trail.

Acknowledgments

Ronald Adkins and Rodney Honeycutt generously provided me with their cytochrome oxidase II (CO II) data. I thank my daughter, Julie, for allowing me to use the computer.

I am very grateful to anonymous reviewers and to Olivier Rieppel and Rodney Honeycutt for helpful criticisms and comments. A very detailed review was provided by Ronald Adkins. Views in this paper are not necessarily in agreement with the readers. Of course, any errors in this paper are my responsibility.

My analyses of certain data summarized herein relied on algorithms provided by the software PAUP (Swofford, 1991b) and HENNIG 86 (Farris, 1988). This research was funded under NSF grant BSR 9106868.

References

Adkins, R. M., and R. L. Honeycutt. 1991. Molecular phylogeny of the superorder archonta. *Proceedings, National Academy of Science*, 88: 1-5.

Ammerman, L. K., and D. M. Hillis. 1990. Relationships within archontan mammals based on 12S rRNA gene sequence.

American Zoologist, 30: 50A.

Bailey, W. J., J. L. Slightom and M. Goodman. 1992. Rejection of the "flying primate" hypothesis by phylogenetic evidence from the e-globin gene. *Science*, 256: 86-89.

Baker, R. J., M. J. Novacek and N. B. Simmons. 1991. On the monophyly of bats. *Systematic Zoology*, 40(2): 216-231.

Beard, K. C. 1990. Gliding behavior and palaeoecology of the alleged primate family Paromomyidae (Mammalia, Dermoptera). *Nature (London)*, 345(6273): 340-341.

Beard, K. C. 1993. Phylogenetic systematics of the Primatomorpha, with special reference to Dermoptera. In F. S. Szalay, M. J. Novacek and M. C. McKenna (Eds.), *Mammal Phylogeny*, Vol. 2. *Placentals*, 129-150. New York: Springer-Verlag.

Brown, W. M., M. George and A. C. Wilson. 1979. Rapid evolution of animal mitochondrial DNA. *Proceedings National Academy of Sciences*, 76: 1967-1971.

Brown, W. M., E. M. Prager, A. Wang and A. C. Wilson. 1982. Mitochondrial DNA sequences of primates: tempo and mode of evolution. *Journal of Molecular Evolution*, 18: 225-239.

Cracraft, J., and K. Helm-Bychowski. 1991. Parsimony and phylogenetic inference using DNA sequences: some methodological strategies. In M. M. Miyamoto and J. Cracraft (Eds.), *Phylogenetic Analysis of DNA Sequences*, 184-220. New York, London: Oxford University Press.

Cronin, J. E., and V. M. Sarich. 1980. Tupaiid and Archonta phylogeny: the macromolecular evidence. In W. P. Luckett (Ed.), *Comparative Biology and Evolutionary Relationships of Tree Shrews*, 293-312. New York, London: Plenum Press.

Czelusniak, J., M. Goodman, B. F. Koop, D. A. Tagle, J. Shoshani, G. Braunitzer, T. K. Kleinschmidt, W. W. De Jong and G. Matsuda. 1990. Perspective from amino acid and nucleotide sequences on cladistic relationships among higher taxa of Eutheria. In H. H. Genoways (Ed.), *Current Mammalogy*, 2: 545-572. New York: Plenum Press.

De Jong, W. W. 1982. Eye lens proteins and vertebrate phylogeny. In

M. Goodman (Ed.), *Macromolecular Sequences in Systematic and Evolutionary Biology*, 75-114. New York, London: Plenum Press.

DeSalle, R., J. Gatesy, W. Wheeler and D. Grimaldi. 1992. DNA sequences from a fossil termite in Oligo-Miocene amber and their phylogenetic implications. *Science*, 257: 1933-1936.

Donoghue, M., J. Doyle, J. Gauthier, A. Kluge and T. Rowe. 1989. The importance of fossils in phylogeny reconstruction. *Annual Reviews of Ecology and Systematics*, 20: 431-460.

Dover, G. A. 1987. Nonhomologous views of a terminology muddle. *Cell*, 51: 525.

Eckenrode, V., J. Arnold and R. Meagher. 1985. Comparison of the nucleotide sequence of soybean 18S rRNA with the sequences of other small subunit rRNAs. *Journal of Molecular Evolution*, 21: 259-269.

Eldredge, N., and M. J. Novacek. 1985. Systematics and paleobiology. *Paleobiology*, 11: 65-74.

Faith, D. P., and P. S. Cranston. 1991. Could a cladogram this short have arisen by chance alone?: on permutation tests for cladistic structure. *Cladistics*, 7: 1-28.

Farris, J. S. 1988. HENNIG 86. Version 1.5. Distributed by the author. 41 Admiral Street, Port Jefferson Station, New York.

Felsenstein, J. 1978. Cases in which parsimony or compatibility methods will be positively misleading. *Systematic Zoology*, 27: 401-410.

Fitch, W. M., and J. Ye. 1991. Weighted parsimony: does it work? In M. M. Miyamoto and J. Cracraft (Eds.), *Phylogenetic Analysis of DNA Sequences*, 147-154. New York, Oxford: Oxford University Press.

Gall, J. G. 1981. Chromosome structure and the c-value paradox. *Journal of Cell Biology*, 91: 3-14.

Gauthier, J., A. Kluge and T. Rowe. 1988. Amniote phylogeny and the importance of fossils. *Cladistics*, 4: 105-209.

Golenberg, E. M., D. E. Giannasi, M. T. Clegg, C. J. Smiley, M. Durbin, D. Henderson and G. Zurawski. 1990. Chloroplast DNA sequence from a Miocene Magnolia species. *Nature (London)*,

344: 656-658.

Goodman, M. 1989. Emerging alliance of phylogenetic systematics and molecular biology: a new age of exploration. In B. Fernholm, K. Bremer and H. Jornvall (Eds.), *The Hierarchy of Life*, 43-61. Nobel Symposium 70. Excerpta Medica. Amsterdam, NY: Elsevier Science Publishers.

Gregory, W. K. 1910. The orders of mammals. *Bulletin, American Museum of Natural History*, 27: 1-524.

Hayasaka, K., T. Gojobori and S. Horai. 1988. Molecular phylogeny and evolution of primate mitochondrial DNA. *Molecular Biology and Evolution*, 5: 626-644.

Hillis, D. M. 1987. Molecular versus morphological approaches to systematics. *Annual Reviews of Ecology and Systematics*, 18: 23-42.

Hillis, D. M. 1991. Discrimination between phylogenetic signal and random noise in DNA sequences. In M. Miyamoto and J. Cracraft (Eds.), *Phylogenetic Analysis of DNA Sequences*, 278-294. New York, Oxford: Oxford University Press.

Hillis, D. M., and M. T. Dixon. 1989. Vertebrate phylogeny: evidence from 28S ribosomal DNA sequences. In B. Fernholm, K. Bremer and H. Jornvall (Eds.), *The Hierarchy of Life*, 355-367. Nobel Symposium 70. Excerpta Medica. Amsterdam, NY: Elsevier Science Publishers.

Holmes, E. C. 1991. Different rates of substitution may produce different phylogenies of the eutherian mammals. *Journal of Molecular Evolution*, 33: 209-215.

Irwin, D. M., T. D. Kocher and A. C. Wilson. 1991. Evolution of the cytochrome b gene of mammals. *Journal of Molecular Evolution*, 32: 128-144.

Kluge, A. G. 1989. A concern for evidence and a phylogenetic hypothesis of relationships among *Epicrates* (Boidae, Serpentes). *Systematic Zoology*, 38: 7-25.

Kraus, F., and M. M. Miyamoto. 1991. Rapid cladogenesis among the pecoran ruminants: evidence from mitochondrial DNA sequences. *Systematic Zoology*, 40(2): 117-130.

Kristensen, N. P. 1981. Phylogeny of insect orders. *Annual Reviews of Entomology*, 26: 135-157.

Lake, J. A. 1987. A rate-independent technique for analysis of nucleic acid sequences: evolutionary parsimony. *Molecular Biology and Evolution*, 4: 167-191.

Lake, J. A. 1988. Origin of the eukaryotic nucleus determined by rate-invariant analysis of rRNA sequences. *Nature (London)*, 331: 184-186.

Lake, J. A. 1989. Origin of the eukaryotic nucleus determined by rate-invariant analysis of ribosomal RNA genes. In B. Fernholm, K. Bremer and H. Jornvall (Eds.), *The Hierarchy of Life*, 87-101. Nobel Symposium 70. Excerpta Medica. Amsterdam, NY: Elsevier Science Publishers.

Li, W.-H., and M. Gouy. 1991. Statistical methods of testing molecular phylogenies. In M. M. Miyamoto and J. Cracraft (Eds.), *Phylogenetic Analysis of DNA Sequences*, 249-277. New York, Oxford: Oxford University Press.

McKenna, M. C. 1975. Toward a phylogenetic classification of the Mammalia. In W. P. Luckett and F. S. Szalay (Eds.), *Phylogeny of the Primates: an Interdisciplinary Approach*, 21-46. New York: Plenum Press.

McKenna, M. C. 1987. Molecular and morphological analysis of high-level mammalian interrelationships. In C. Patterson (Ed.), *Molecules and Morphology in Evolution: Conflict or Compromise?*, 55-93. Cambridge: Cambridge University Press.

McKenna, M. C. 1992. The alpha crystallin A chain of the eye lens and mammalian phylogeny. *Annales Zoologici Fennici*, 28: 349-360.

MacFadden, B. J. 1992. Interpreting extinctions from the fossil record: methods, assumptions, and case examples using horses (Family Equidae). In M. J. Novacek and Q. D. Wheeler (Eds.), *Extinction and Phylogeny*, 17-45. New York: Columbia University Press.

MacPhee, R. D. E., M. J. Novacek and G. Storch. 1988. Basicranial morphology of early Tertiary erinaceomorphs and the origin of primates. *American Museum Novitates*, 2921: 1-42.

Maddison, W. P., M. J. Donoghue and D. R. Maddison. 1984. Outgroup analysis and parsimony. *Systematic Zoology*, 33: 83-103.

Mindell, D. P. 1991. Aligning DNA sequences: homology and phylogenetic weighting. In M. M. Miyamoto and J. Cracraft (Eds.), *Phylogenetic Analysis of DNA Sequences*, 73-89. New York, Oxford: Oxford University Press.

Mindell, D. P., C. W. Dick and R. J. Baker. 1991. Phylogenetic relationships among megabats, microbats, and primates. *Proceedings National Academy of Science*, 88: 10322-10326.

Miyamoto, M. M., and S. M. Boyle. 1989. The potential importance of mitochondrial DNA sequence data to eutherian mammal phylogeny. In B. Fernholm, K. Bremer and H. Jornvall (Eds.), *The Hierarchy of Life*, 437-452. Nobel Symposium 70. Excerpta Medica. Amsterdam, NY: Elsevier Science Publishers.

Miyamoto, M. M., and M. Goodman. 1986. Biomolecular systematics of eutherian mammals: phylogenetic patterns and classification. *Systematic Zoology*, 35: 230-240.

Nei, M. 1991. Relative efficiencies of different tree-making methods for molecular data. In M. M. Miyamoto and J. Cracraft (Eds.), *Phylogenetic Analysis of DNA Sequences*, 90-128. New York, Oxford: Oxford University Press.

Norell, M. A., and M. J. Novacek. 1992. The fossil record and evolution: comparing cladistic and paleontologic evidence for vertebrate history. *Science*, 255: 1690-1693.

Novacek, M. J. 1982. Information for molecular studies from anatomical and fossil evidence on higher eutherian phylogeny. In M. Goodman (Ed.), *Macromolecular Sequences in Systematic and Evolutionary Biology*, 3-41. New York: Plenum Press.

Novacek, M. J. 1986. The skull of leptictid insectivorans and the higher-level classification of eutherian mammals. *Bulletin, American Museum of Natural History*, 183: 1-111.

Novacek, M. J. 1990. Morphology, paleontology, and the higher clades of mammals. In H. H. Genoways (Ed.), *Current Mammalogy*, 2: 507-543. New York: Plenum Press.

Novacek, M. J. 1992. Fossils as critical data for phylogeny. In M. J.

Novacek and Q. D. Wheeler (Eds.), *Extinction and Phylogeny*, 46-88. New York: Columbia University Press.

Novacek, M. J., A. Wyss and M. C. McKenna. 1988. The major groups of eutherian mammals. In M. Benton (Ed.), *The Phylogeny and Classification of the Tetrapods*. Vol. 2. *Mammals*, 2: 31-71. Oxford: Clarendon Press.

Patterson, C. (Ed.). 1987. *Molecules and Morphology in Evolution: Conflict or Compromise?*, 229 pp. Cambridge, London: Cambridge University Press.

Patterson, C. 1989. Phylogenetic relations of major groups: conclusions and prospects. In B. Fernholm, K. Bremer and H. Jornvall (Eds.), *The Hierarchy of Life*, 471-488. Nobel Symposium 70. Excerpta Medica. Amsterdam, NY: Elsevier Science Publishers.

Pettigrew, J. D. 1986. Flying primates? Megabats have the advanced pathway from eye to midbrain. *Science*, 231: 1304-1306.

Pettigrew, J. D. 1991. Wings or brain? Convergent evolution in the origins of bats. *Systematic Zoology*, 40: 199-216.

Platnick, N. I., C. E. Griswold and J. A. Coddington. 1991. On missing entries in cladistic analysis. *Cladistics*, 7: 337-343.

Raff, R. A., K. G. Field, G. J. Olsen, S. J. Giovannoni, D. J. Lane, M., T. Ghiselin, N. R. Pace and E. C. Raff. 1989. Summary of our present knowledge of metazoan phylogeny. In B. Fernholm, K. Bremer and H. Jornvall (Eds.), *The Hierarchy of Life*, 261-272. Nobel Symposium 70. Excerpta Medica. Amsterdam, NY: Elsevier Science Publishers.

Reeck, G. R., C. de Häen and D. C. Teller. 1987. "Homology" in proteins and nucleic acids: terminology muddle and a way out of it. *Cell*, 50: 667.

Sanderson, M., and M. Donoghue. 1989. Patterns of variation in levels of homoplasy. *Evolution*, 43: 1781-1795.

Sidow, A., and A. C. Wilson. 1991. Compositional statistics evaluated by computer simulation. In M. Miyamoto and J. Cracraft (Eds.), *Phylogenetic Analysis of DNA Sequences*, 129-146. New York, Oxford: Oxford University Press.

Simmons, N., M. J. Novacek and R. J. Baker. 1991. Approaches, methods, and the future of the chiropteran monophyly controversy: a reply to J. D. Pettigrew. *Systematic Zoology*, 40: 239-243.

Simpson, G. G. 1975. Recent advances in methods of phylogenetic inferences. In P. Luckett, and E. Delson (Eds.), *Phylogeny of the Primates*, 3-19. New York: Plenum Press.

Smith, J. D., and G. Madkour. 1980. Penial morphology and the question of chiropteran phylogeny. In D. E. Wilson and A. L. Gardner (Eds.), *Proceedings, 5th International Bat Research Conference*, 347-365. Lubbock: Texas Tech Press.

Sogin, M. L., J. H. Gunderson, H. J. Elwood, R. A. Alonso and D. A. Peattie. 1989. Phylogenetic meaning of the kingdom concept: An unusual ribosomal RNA from *Giardia lamblia*. *Science*, 243: 75-77.

Swofford, D. L. 1991a. When are phylogeny estimates from molecular and morphological data incongruent? In M. M. Miyamoto and J. Cracraft (Eds.), *Phylogenetic Analysis of DNA sequences*, 295-333. New York, Oxford: Oxford University Press.

Swofford, D. L. 1991b. PAUP. Phylogenetic analysis using parsimony. User Manual, version 3.0. Champaign, IL: Illinois Natural History Survey.

Szalay, F. S., and W. J. Bock. 1991. Evolutionary theory and systematics: relationships between process and patterns. *Zeitschrift Zoologische Systematik und Evolutions-forschung*, 29: 1-39.

Thewissen, J. G., and S. K. Babcock. 1991. Distinctive cranial and cervical innervation of wing muscles: new evidence for bat monophyly. *Science*, 251: 934-936.

Thiele, A., M. Vogelsang and K.-P. Hoffmann. 1991. Patterns of retinotectal projection in the megachiropteran bat *Rousettus aegyptiacus*. *Journal of Comparative Neurology*, 314: 671-683.

Vrana, P. (in prep.) Molecular systematic studies of the placement of Carnivora within eutherian mammals.

Waterman, M. S., J. Joyce and M. Eggert. 1991. Computer alignment of sequences. In M. M. Miyamoto and J. Cracraft (Eds.),

Phylogenetic Analysis of DNA Sequences, 59-72. New York, Oxford: Oxford University Press.

Wheeler, Q. D. 1990. Ontogeny and character phylogeny. *Cladistics*, 6: 225-268.

Wheeler, W. C. 1989. The systematics of insect ribosomal DNA. In B. Fernholm, K. Bremer and J. Jornvall (Eds.), *The Hierarchy of Life*, 307-324. Nobel Symposium 70. Excerpta Medica. Amsterdam, NY: Elsevier Science Publishers.

Wheeler, W. C. 1990. Combinatorial weights in phylogenetic analysis: a statistical parsimony procedure. *Cladistics*, 6: 269-275.

Wheeler, W. C. 1992. Extinction, sampling, and molecular phylogenetics. In M. J. Novacek and Q. D. Wheeler (Eds.), *Extinction and Phylogeny*, 205-215. New York: Columbia University Press.

Wheeler, W. C., and D. M. Gladstein. 1992. Malign. Program and Documentation. Version 1.01. Distributed by author. American Museum of Natural History, New York, NY.

Wheeler, W. C., and R. L. Honeycutt. 1988. Paired sequence difference in ribosomal RNAs: evolutionary and phylogenetic implications. *Molecular Biology and Evolution*, 5: 90-96.

Wible, J. R., and M. J. Novacek. 1988. Cranial evidence for the monophyletic origin of bats. *American Museum Novitates*, 2911: 1-19.

Williams P. L., and W. M. Fitch. 1989. Finding the minimal change in a given tree. In B. Fernholm, K. Bremer and H. Jornvall (Eds.), *The Hierarchy of Life*, 453-470. Nobel Symposium 70, Excerpta Medica. Amsterdam, NY: Elsevier Science Publishers.

Woese, C. R. 1989. Archaebacteria and the nature of their evolution. In B. Fernholm, K. Bremer and H. Jornvall (Eds.), *The Hierarchy of Life*, 119-132. Nobel Symposium 70. Excerpta Medica. Amsterdam, NY: Elsevier Science Publishers.

Wyss, A. R., M. J. Novacek and M. C. McKenna. 1987. Amino acid sequence versus morphological data and the interordinal relationships of mammals. *Molecular Biology and Evolution*, 4: 99-116.

Zimmer, E. A., R. K. Hamby, M. L. Arnold, D. A. LeBlanc and E. C. Theriot. 1989. Ribosomal RNA phylogenies and flowering plant evolution. In B. Fernholm, K. Bremer and H. Jornvall (Eds.), *The Hierarchy of Life*, 205-214. Nobel Symposium 70. Excerpta Medica. Amsterdam, NY: Elsevier Science Publishers.

6

Stratocladistics: Morphological and Temporal Patterns and their Relation to Phylogenetic Process

Daniel C. Fisher

Museum of Paleontology and Department of Geological Sciences
University of Michigan
Ann Arbor, Michigan 48109

Abstract. Patterns of distribution of morphological or molecular synapomorphies among taxa can be interpreted as a hierarchy of sister group relationships without explicit reference to time. However, sister group relations are explained by descent from a common ancestor, and the differential distribution of synapomorphies can be taken as evidence of the order of branching of descendant lineages. Both descent and branching are time-dependent processes and imply an ordering of taxa in time. Furthermore, competing hypotheses of relationship imply different temporal orderings of taxa and thus different expected temporal patterns of occurrence. The observed temporal pattern of occurrence of taxa may thus differentially support competing hypotheses of relationship as well as providing constraints on the timing of phylogenetic events. Nontrivial process-assumptions are required for interpreting synapomorphy distributions and temporal ordering, but use of temporal data transfers much of the inferential burden to assumptions that can be evaluated on stratigraphic and taphonomic grounds–i.e., outside the domain of evolutionary biology.

 Copyright © 1994 by Academic Press, Inc.
All rights of reproduction in any form reserved.

ISSUES OF PATTERN, PROCESS, AND THEIR INTERRELATION are fundamental to the whole field of systematic biology. They affect how we design our investigations, why we undertake them, and what we presume to make of the results. These issues lie at a deep enough level that practicing systematists often proceed on an operational plane, as if the underlying rationale had long before taken on a stable configuration, but as we all know, we are still tinkering with the pieces, even making major readjustments to the foundation, while work proceeds above. For science, of course, this is business as usual.

In particular, the nature and role of process assumptions in the perception and analysis of phylogenetic patterns have elicited strong and frequently divergent opinions, to the point that these issues define one of the major philosophical watersheds evident on the theoretical landscape of systematic biology today (Sober, 1988). When differences of practice and perspective run so deep, it can be difficult to find a path of reconciliation, but that is what I am looking for in the lines of thought described below. I am not interested in reconciliation for its own sake, at the cost of intellectual compromise, but I do want to discover whether there might in fact be some one, consistent, and valid perspective that would unify the insights that I think lie at the core of each of the presently opposed views.

At one level, this discussion focuses on two schools of thought described most simply as "pattern cladistics" (Beatty, 1982) and "phylogenetic systematics" (*sensu* de Queiroz and Donoghue, 1990). Any convenient summary grossly oversimplifies, but pattern cladists tend to characterize their approach to documenting patterns as relying minimally on evolutionary process assumptions. In contrast, phylogenetic systematists tend to acknowledge deeper reliance on assumptions about characters, their evolutionary independence, and their comparative evidential value, although how these assumptions translate into substantive statements about evolutionary process remains a matter of considerable controversy (Sober, 1988). As implied here, the differences between schools may be more in how we represent what we do, than in what we actually do. Neither pattern cladistics nor phylogenetic systematics is a monolithic assemblage. The issues that divide our field are so many and so complex that they allow no simple dichotomy of practitioners. Nevertheless, "schools of thought" are useful for describing sets of ideas, even if their application to real people is fuzzy.

At another level, this discussion contrasts pattern cladistics with a more paleontological perspective on documenting life's pattern. I refer here to no identifiable school of paleontological practice, but rather to the more general proposition that stratigraphic data, information about the temporal ordering of occurrences of fossils, might be relevant to identifying phylogenetic pattern. Of course, paleontologists represent a wide variety of approaches to the discovery of phylogenetic pattern, and not all see merit in using data on temporal ordering to choose among alternative cladistic hypotheses. Nevertheless, data on temporal ordering are inherently paleontological, and paleontologists include in their number some who do use temporal data (e.g., Gingerich, 1979). I am aware that some readers will think of this as a battle fought and won long ago by cladistics, leaving temporal data only a minor role in the systematic enterprise, such as in estimating times of divergence of taxa identified on other grounds. I beg the indulgence of such readers, however, for I believe this issue deserves reconsideration, and I will propose a substantially different approach to using temporal data.

My exploration of these issues will focus on several questions discussed at length in previous literature (e.g., Eldredge and Cracraft, 1980; Beatty, 1982; Brady, 1985) and in other contributions to this volume.

> 1. How, in general, are pattern and process related? I will treat this at the beginning, partly to clarify different contexts in which these terms will be used, but more importantly because some aspects of the relation between pattern and process seem so general that they may be able to inform us about pattern/process interaction in systematic biology.

2. What can pattern cladistics do by itself? That is, does pattern cladistics, independent of assumptions about evolutionary process, provide a sufficient basis for a description of the "order" that we observe among organisms?

3. What was the basis for pre-Darwinian perceptions of Life's "order"? As Brady (1985; Chapter 2, this volume) has reminded us, there were, prior to Darwin, well articulated notions of hierarchical organization of the diversity of organisms, framed under no commitment to ideas of descent, branching, or evolutionary transformation (as we would use the phrase), let alone natural selection or any other process of evolutionary change.

4. How can we, operating now in a context where evolutionary thinking has broad currency, resolve patterns of organization among organisms with minimal reliance on assumptions about evolutionary process? This, clearly, is the aim of pattern cladists, and I accept it as an expression of concern for the validity of inferences, not a reflection of antagonism toward evolutionary theory.

My treatment of these questions will take a number of twists and turns, here and there following, but never more than a short distance, the paths trod by others in this volume. Moreover, by the end of this discussion, I will have suggested a union of cladistics and stratigraphic paleontology that will seem to some a most unholy matrimony, but I believe it holds the answer to what pattern cladists have been seeking all along.

PATTERN AND PROCESS

The most important element of pattern is repeatability. This usually implies at least inter-observer replicability, but also usually involves observation of similar structure inherent in what seem to be different instances of the "same" phenomenon. The intent of notions of process is to explain this repeated structure in terms of a cause or causes acting in the system within which pattern arises. To talk of process is to postulate some time-dependent behavior for the system in which pattern is observed. In some cases, time is already one of the coordinates used in the description of pattern, but *if* it is not, and I will deal here with cases where it is not, speaking of process implicitly brings time into the picture. In any event, pattern and process are related as data and hypothesis (Fig. 1). That is, we are interested in patterns as

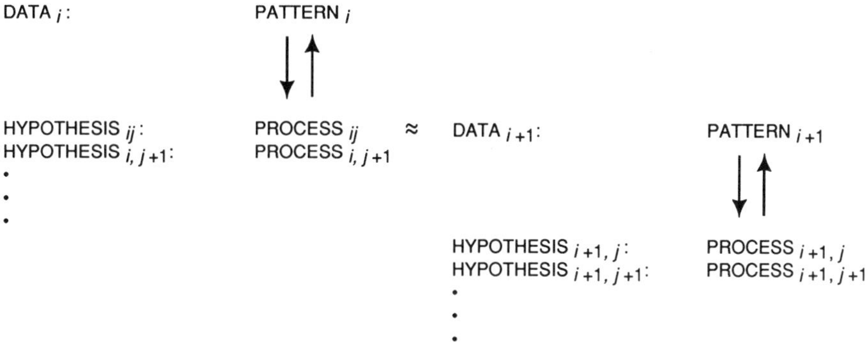

Figure 1. Diagram of the hierarchical relation between pattern and process. Corresponding pattern and process statements (same value of first subscript, i) are consistently displaced by a level in the analytical hierarchy. Any one pattern (or set of data) may be explained by multiple (consecutive values of second subscript, $j, j + 1,...$) competing process-hypotheses. Arrows reflect the *explanans/explanandum* relation discussed by Brady (1985).

evidence for evaluating competing hypotheses about process, but we are interested in processes because they make statements about relations, or dimensions, that were not explicit in the original pattern data. This is the source of the explanatory power of process hypotheses (Brady, 1985). Cautious of this power, we demand some degree of independence between pattern description and hypothesis evaluation, but in acknowledging the theory-bound nature of observation, we recognize that this independence is not absolute. Only by recognizing exactly which hypotheses of process affect our perception of pattern, and in what way, may we avoid vicious circularity (Hull, 1967).

As an example of this data versus hypothesis relation between pattern and process, consider a trackway–a series of footprints–on a sandy substrate. This is a pattern with repeated elements at several scales, the most obvious of which is the repetition of footprints along the series. Observing this pattern gives evidence for evaluating alternative causal hypotheses, any one of which refers to a process by which the trackway might have been made. Alternative process hypotheses may vary, for instance, in the presumed identity of the trackmaker and speed of locomotion. Entertaining any such hypothesis has the power to "guide the eye" in following a pattern thought to be a trackway, and hence we reserve this description for cases where elements of the pattern (footprints, or stride sequences) can be evaluated independently and satisfy standard criteria of recognition (e.g., shape and local topographic relief relative to other substrate irregularities) whose application is not redundant to observation of the higher order pattern of repetition that comprises the trackway itself.

The aspect of pattern and process that is most frequently overlooked is their hierarchical relation and the consequent relativity of many assignments of terms to pattern or process categories (Fisher, 1985). In other words, identification of a term as designating either a pattern or a process reflects an implied relation to some second term above or below the first within a hierarchy of data and hypotheses. In much of our thinking and talking, we move up and down through the levels of this hierarchy, changing our reference point subtly as we go. The trackway introduced above represents a pattern that can be described topographically in a completely atemporal fashion. Analyzing its data, we may decide that the process responsible for it was a quadrupedal vertebrate walking. Yet vertebrate walking is also a *pattern* of locomotion, defined by various anatomical features and lack of a suspended phase–a pattern of interaction between an organism and a substrate. This pattern is itself data for investigating underlying process, which, among other things, turns out to involve muscle contraction. But wait, muscle contraction is also a pattern, a pattern of displacement of tissues driven by a process...and so on into the minutiae of physiology and biochemistry. We routinely recognize that patterns are used to investigate processes, but a statement of *process* at one level of analysis expresses a relationship that may then be treated as a *pattern* for purposes of a finer scale inquiry into process. Corroborated hypotheses of process *become the patterns* used as data for a deeper investigation of process. It is thus not inconsistent to refer to descent, branching, or transformation as phylogenetic *processes*, and then later refer to them as *patterns* used in investigating evolutionary processes such as selection or drift. "Phylogenetic" and "evolutionary" here identify levels of analysis, the former referring to the deployment of lineages through time (whether or not time is measured in absolute units), and the latter referring to interactions within lineages.

A further complication is that when we consider certain process hypotheses unproblematic, we sometimes use them to integrate simple patterns into more complex patterns. The trackway discussed above can be studied either at the scale of individual footprints (relatively simple patterns) or an ensemble of footprints (a more complex pattern involving aspects of spacing and orientation not observable within a single print). The ensemble *could* be described in a process-neutral fashion, but calling it a trackway is not process-neutral because this implies a sequential relation (we observe on the substrate a spatial array, not a temporal sequence) between the elements of the ensemble and identifies the ensemble as a record of locomotor performance. Yet it is only with reference to a process that we consider the ensemble a "natural group" with its own intelligible structure. A competing process hypothesis, such as that the ensemble of depressions was formed by a large template pressed onto the substrate in a single event, would imply different relations between individual elements of the pattern and would be incompatible with calling the ensemble a trackway. When this integrative use of process hypotheses introduces information regarding a dimension, such as time, to which we do not have direct access in a given case, it may be said to have increased the dimensionality of the pattern. Time is not the only dimension that can be added, but it is commonly introduced in this way. Based on these observations, I react with doubt when I hear a claim that one can take pattern data and deduce a more highly dimensional pattern without benefit of process assumptions. It sounds like getting something for nothing, and in my experience that is even less common in inference than it is in economics.

PATTERN CLADISTICS

The fundamental claim of at least "hardline" pattern cladistics is that a pattern of character state distributions, evaluated only by searching for congruence, implies a unique taxonomic hierarchy, or classification– groups nested neatly within groups (Fig. 2). The type of congruence involved here is the kind of "agreement" among patterns of character state distribution that allows them to be mapped onto the same cladogram (Patterson, 1982; Nelson, 1989). The cladogram in Figure 2 is taken simply as an expression of this grouping of taxa, or as a generalization about the distribution of character states, and is not held to imply anything about descent. The hierarchical pattern expressed by this cladogram is supposedly derived from character data *without benefit of evolutionary process assumptions*. Once in hand, the order expressed in that hierarchy stands as something that may be explained, as indicated in Figure 3, by hypotheses of descent, branching, and transformation, and the degree to which it is so explained becomes part of the basis for judging those hypotheses. But there is, according to this representation, no participation of notions of phylogenetic or evolutionary process (as distinguished above) in the translation of character distributions into the taxonomic hierarchy.

In Figure 3, character distributions are described as two-

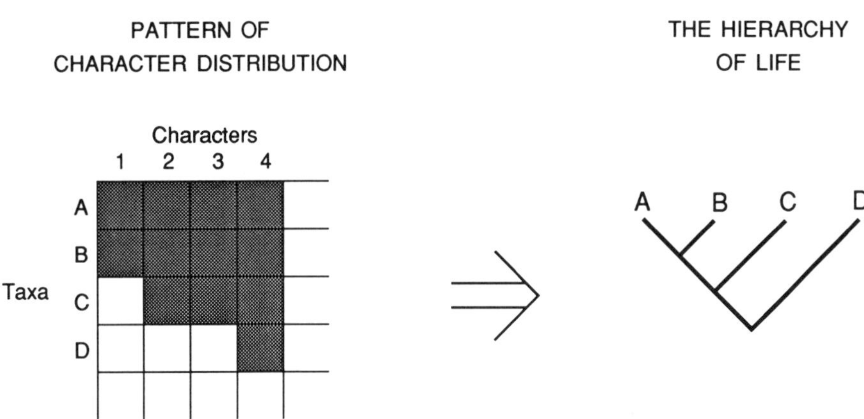

Figure 2. Diagram of the central claim of pattern cladistics, that the pattern of character distribution implies a unique hierarchy among organisms. The taxon x character matrix is "open" in both dimensions, indicating that the pattern displayed is representative, not exhaustive.

dimensional; of course characters are of various dimensionalities, but their *distribution* in taxon x character matrices may be treated as a two-dimensional pattern. Whatever dimensionality we assign such a matrix, there is clearly "something extra" in the taxonomic hierarchy, for it specifies a pattern of grouping that is not itself observable *as an attribute of organisms* and is thus not coded directly in the taxon x character matrix. This is indicated in Figure 3 by assigning subscript 1.5 (i.e., intermediate between $Pattern_1$ and $Pattern_2$) to the pattern represented by the taxonomic hierarchy and by referring to the latter as three-dimensional. So far, the pattern cladist claim looks like an advertisement for an epistemological free lunch.

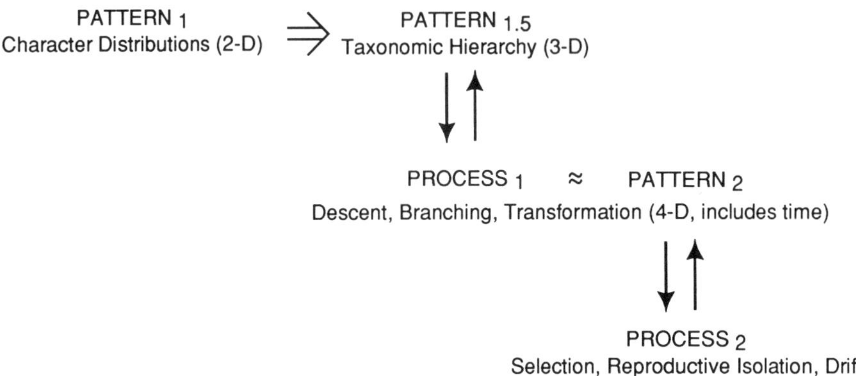

Figure 3. Diagram (following the schema of Fig. 1) of the relation, according to pattern cladistics, between classification (upper statement of entailment, from Fig. 2) and process-oriented aspects of systematic and evolutionary biology. Process_1 and Pattern_2 refer to phylogenetic phenomena, those at the level of lineages (e.g., descent, branching, transformation), and Process_2 refers to evolutionary phenomena, among individuals or groups of individuals, within lineages (e.g., selection, reproductive isolation, drift).

In deference to the full range of views associated with pattern cladism, not all its representatives agree that congruence is enough. This stance is discussed further below, but even it contrasts with phylogenetic systematics (*sensu* de Queiroz and Donoghue, 1990), which readily acknowledges reliance on the hypothesis of "descent with modification" and thus recognizes the integrative role of a hypothesis about phylogenetic process in generating hierarchies from character distributions. As noted in the introduction, the question faced by "untransformed" cladists is not *whether* process hypotheses participate in perceptions of the taxonomic hierarchy but rather *which* process hypotheses, from *which* levels of analysis (phylogenetic and/or evolutionary).

DOES PATTERN CLADISTICS WORK?

Figure 4 looks more closely at the operation of pattern cladistics. As it happens, observations do not come, already coded, in matrices. They come in limbs and tissues and behaviors. Individuation of such traits, and evaluation of their independence or redundancy, require nontrivial judgments on issues with which evolution deals, for instance, how the diversity of character states we observe arose. Furthermore, the character observations that follow from those judgments are not always internally consistent (indicated by a departure from congruence in the uppermost taxon x character matrix), and sometimes are susceptible to alternative representation (indicated by the open set of alternative taxon

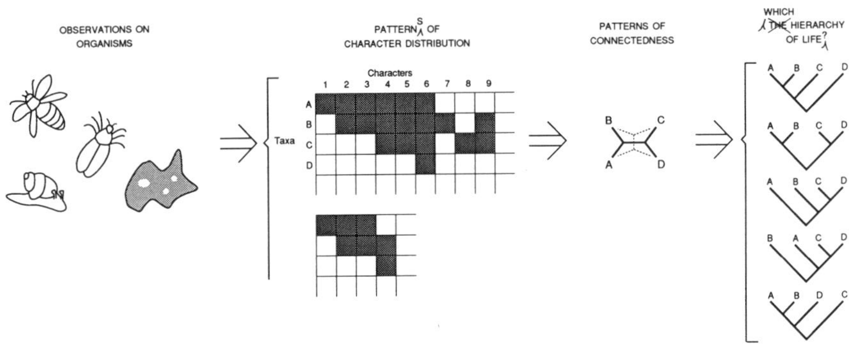

Figure 4. Expanded and revised diagram of the operation of pattern cladistics (cf. Fig. 2). Unbiased abstraction of characters from real taxa is nontrivial, data sets generally show imperfect congruence (character 9 versus others in the uppermost taxon x character matrix), and observations may be interpreted in different ways (hence the open set of taxon x character matrices, the second of which differs from the first). In many cases, a single, unrooted network may still best summarize the pattern of most observations (solid lines show this network; dotted lines show an alternative pattern, just behind the first in degree of support), but even in this "best case" scenario, each alternative rooting of the unrooted network yields a competing hierarchy of equal merit by the criterion of congruence.

x character matrices for the same set of specimens, or biological individuals). However, these points are not new, nor in the end are they as critical as some others.

A more interesting problem arises when we realize that focusing on even a single, relatively congruent pattern of character distribution does not give, by itself, a unique grouping of taxa. For the untransformed cladists among us, this becomes apparent when we remember that parsimony, our operational criterion of congruence (Farris, 1983; Sober, 1988; Wiley et al., 1991), suffices to select an *un*rooted network from among the large number of graphs that might a priori connect a series of ingroup taxa, but we need something more to root that network and turn it into a cladogram expressing hierarchical order (Lundberg, 1972; Meacham, 1984). A pattern cladist would not explain it just this way, because "connectedness" might (though it need not) betray some notion of descent, but the problem still remains. Congruence greatly pares down the initial universe of possibilities, but there is always a nontrivial residuum among which it is powerless to choose.

Before cries of "Outgroup! Outgroup! Bring on the outgroup!" (or "Ontogeny!...") swell in readers' throats, let me explore this point a bit further (Fig. 5). For a three-taxon problem there is only one unrooted network and three dichotomous cladograms. Suppose we are dealing with independent, binary characters and have, for the moment, no outgroup (or ontogeny) available. We can make no appeal to general and special, primitive and derived, only recognize alternative conditions. There are only four ways, each indicated by a row of cells above each cladogram in Figure 5, in which such states can be distributed among three taxa. Remember, black and white are interchangeable; only the contrast records the pattern of observations. Thus, there is no necessary consistency from row to row in the significance of a given color, and any possible pattern of distribution can be represented as one of these four rows. The four rows are not to be taken as representing a complete character matrix; they only show the patterns of state distribution possible. Uniform character state distributions, such as on the bottom row, are compatible with any of the cladograms and thus do not choose among them. The only basis for selecting the cladogram on the left would be a relative abundance of

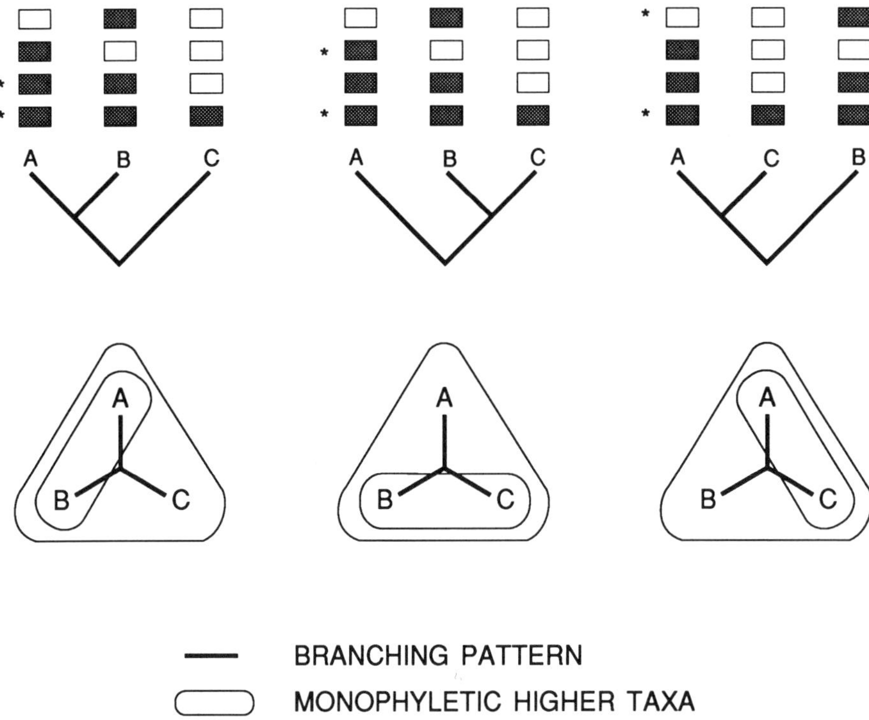

Figure 5. Cladogram selection for three taxa. Only one unrooted network is possible, and three dichotomous cladograms. Any possible pattern of character state distribution (rows of cells above each cladogram) is compatible with at least one of these cladograms (i.e., each row might be expected to represent a frequently observed pattern given the cladogram for which it bears an asterisk). The monophyletic groups implied by each cladogram are indicated by solid boundaries on the unrooted network below it. Different frequencies of observation of character patterns may suggest one cladogram over others, but this amounts to nothing more than a phenetic analysis.

characters showing the distribution of the third row of cells (marked there with an asterisk), for this represents characters shown only by A and B, and this cladogram groups those two together, as indicated also on the network, by a solid boundary enclosing A and B, apart from C. At first this may seem to solve the problem. If characters showing this distribution outnumber characters distributed as in the first two rows, didn't I just say this was the supported hierarchy? Well, yes, but then our final choice would have been based on nothing more than a *phenetic* analysis, grouping taxa by overall similarity. And the problem doesn't get easier when there are more taxa (Fig. 6). I need not retrace all the steps here; the result is similar. In this case, congruence, by virtue of selecting just one of the possible unrooted networks (only three are possible for four taxa, assuming we are interested only in dichotomous cladograms, but this number increases rapidly with larger numbers of taxa) lets us focus on a proper subset of all possible cladograms. Thus, it does carry information. It reflects a general paucity of characters with the distributions represented by the sixth and seventh rows of cells above cladograms in Figure 6. But for *choosing among* the cladograms that are left, none of the other patterns of character distribution operates cladistically, because each can be interpreted in a way that is compatible with any of the cladograms shown. Attempts to select among these cladograms degenerate, as in Figure 5, to a phenetic exercise.

Thus far, I have focused on congruence, but *in*congruence is also important. The closed boundaries overlaid on each network diagram in Figure 6 constitute a Venn diagram (Wiley et al., 1991) representing taxa recognized by the cladogram just above, and incongruence would show up as crossed boundaries or partial overlap of regions on the Venn diagram. The two character patterns not expected under any of the cladograms in Figure 6 are resemblances between A and C on one hand, B and D on the other, or between A and D on one hand, B and C on the other. A closed curve representing such a pairing would be incongruent with groups recognized in Figure 6. For instance, moving from left to right through the series of five unrooted network diagrams, grouping by a state shared by A and C would intersect two taxonomic

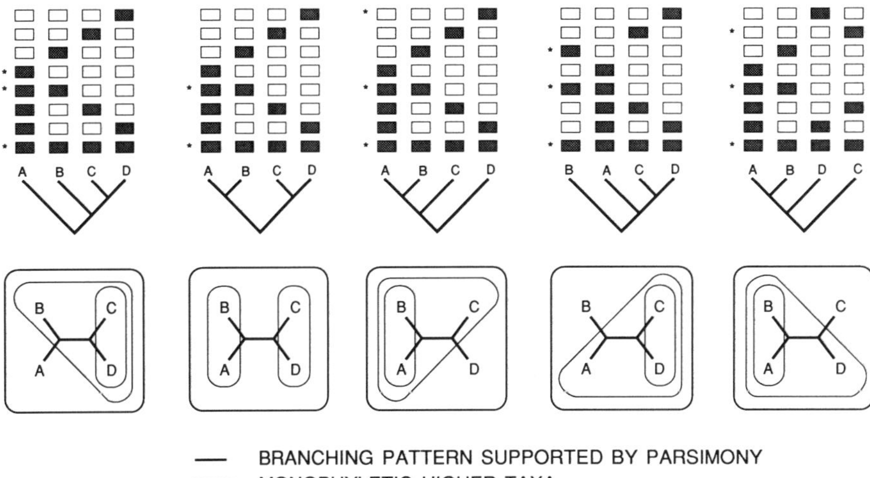

Figure 6. Cladogram selection for four taxa. Congruence may now select one (shown on the lower portion of the diagram) of the three possible unrooted networks (for which no more than three links join at each interior node), and this in turn is compatible with five dichotomous cladograms (shown above). There are now eight possible patterns of character state distribution (rows of cells above each cladogram), of which six are compatible with one or more of these cladograms. Each of these six might be expected to represent a frequently observed pattern given the cladogram for which it bears an asterisk. The sixth and seventh rows of cells, counting from the top, represent character patterns that are not expected under any cladogram associated with this unrooted network (they would each have supported an alternative unrooted network). The monophyletic groups implied by each cladogram are indicated by solid boundaries on the unrooted network below it. As in Figure 5, different frequencies of observation of character patterns may suggest one cladogram over others, but selection among the cladograms shown here is essentially a phenetic analysis.

curves, then two again, then one, then one again, and finally two.

If systematics truly operates only on congruence, and yet congruence fails to resolve a unique hierarchy, why has the systematics enterprise not ground to a halt long ago? The answer is that people bring to their analyses a variety of sorting criteria beyond congruence. One such criterion, said to comprise a fundamental distinction between phenetics and cladistics, is that "cladistics operates by grouping on the basis of presence, not absence" (C. Patterson, pers. commun., 1992). The distinction between presence and absence is given hierarchical content by stating that "presence is always a subset of absence" (C. Patterson, pers. commun., 1992). It is easy enough to see how this approach might pick out a single hierarchy: for any character, some but not all cladograms associated with a given unrooted network allow presence to be a subset of absence, and treatment of additional characters with different distributions tends to erode the membership of the intersection set of cladograms consistent with this set relation. In fact, if we assume equivalence between grouping by presence and grouping by synapomorphies, it may appear that we can directly construct most-congruent cladograms, component by component, without ever worrying about unrooted networks (Nelson and Platnick, 1981). However, this simply restates the order of application of criteria; presence as a subset of absence is still a criterion separate from congruence.

What other implications does the presence/absence criterion carry? If it were viewed within the context of an evolutionary model, it would mean that evolution always proceeds by adding features, never by losing them, but such an assertion would be far from the spirit of pattern cladism. Perhaps the "purest" pattern cladistic response is to argue that making presence a subset of absence is at least a criterion that can be applied across the board and that has the effect of drawing attention to a single hierarchical grouping and to suggest we should be satisfied with this if our intent is simply to discover pattern in nature. One problem with this response is that many characters show alternative states neither one of which is an unambiguous instance of presence or absence (e.g., horn shape: coiled or straight). We can always transform these into cases of presence and absence (e.g., presence or absence of coiled form on the horns, or is it presence or absence of straight form?), but doing so is just an exercise in verbal ingenuity, and the cladistic resolution achieved seems quite arbitrary. Another problem is that grouping by *absence* would do as well (or as poorly) if all we are interested in is "pure pattern." A possible response to this is that "presence is a subset of absence" is an objective law of comparative morphology, but this is equivalent to saying (*contra* my above-stated intent) that the information in the taxon x character matrices in Figures 5 and 6 is "clearly" in the shaded cells as contrasted with the fundamentally white background of all cells. The art of M. C. Escher should have taught us the fallacy of this position long ago. Finally, it may be argued that making presence a subset of absence is the operative criterion in systematics because this relation is observed in ontogeny. This and related matters (such as using ontogeny to identify synapomorphies even when not couched in terms of presence/absence) have been discussed at length recently (e.g., Nelson, 1978; Rieppel, 1985, 1990) and are very controversial (de Queiroz, 1985; Kluge, 1985, 1988; Mabee, 1989). While I cannot do justice to the problem within the space constraints of this chapter, it seems that there are only two ways to proceed. Either we combine data from character distributions and ontogenetic transformations without justification, yielding an arbitrary resolution of the taxonomic hierarchy, or we address up front the relevance of ontogeny to the goal of grouping organisms. This might be done by appealing to von Baer's law or to some other generalization–I do not want to limit our options here–but one way or another, there is a gulf to be bridged. Without some notion of process to make the connection, it is not clear how ontogenetic transformation within individuals relates to the diversity of character states observed in taxa, two levels away in the analytical hierarchy diagrammed in Figure 3. Unless we are satisfied with an arbitrary resolution, we need a process hypothesis to integrate ontogenetic patterns with character distribution patterns to yield a taxonomic hierarchy.

A less "hardline" response to the shortfall of congruence alone is to acknowledge that it selects only unrooted networks but argue that its operation cannot be understood in terms of a single "round" of analysis; that is, all analyses of taxonomic order take place within a context of prior analyses which allow us to identify outgroups. Introduction of outgroups will of course "root the tree," but this involves putting phylogenetic information, or at a minimum, classificatory information (if one insists that outgroups reflect only grouping, rather than any statement about relative recency of common ancestry), *in*to the analysis. I do not suggest that this makes cladistics viciously circular. Vicious circularity, or tautology, would involve putting in the *same* information that we presume to get out whereas use of outgroups only involves putting in the same *kind* of information we presume to get out (Sober, 1988). This distinction is critical for understanding "reciprocal illumination" (Hennig, 1966), the significance of levels of analysis (Hull, 1967), and appeals to "research cycles" (Kluge, 1991). Still, reliance on outgroups raises the question of where the buck stops, as we expand our purview, and well this side of any infinite regress, we have to wrestle with those intermediate-level groups for which we simply have not been able to resolve an informative outgroup. In certain cases, ordered multistate characters or so-called "safe" assumptions likc "simplc to complex" will suffice to root a cladogram, but let's be hardnosed and fully transformed about this. When all is said and done, can we justify a unique hierarchy without introducing classificatory, phylogenetic, or other evolutionary information?

Although pattern cladists have used the existence of pre-Darwinian hierarchical classifications to suggest that notions of evolution are not required for demonstrating hierarchical order (Brady, 1985), neither ontogeny and congruence, nor outgroups and congruence, seem likely to have been the interacting criteria responsible for most pre-Darwinian classifications. I have undertaken no historical inquiry along these lines myself, but when I read Darwin's comment that some naturalists look on the Natural System "as a scheme for arranging together those living objects which are most alike, and for separating those which are most unlike" (Darwin, 1859, p. 413), an obvious possibility is suggested, that the Natural System, however much influenced by perceptions of congruence, was also an expression of phenetic relationships. As noted above, a phenetic analysis (assuming agreement on how to conduct it), overlaid on the results of a search for congruence, would suffice to select a unique hierarchy. This proposal is made more explicit in Figure 7, where the cladograms and unrooted

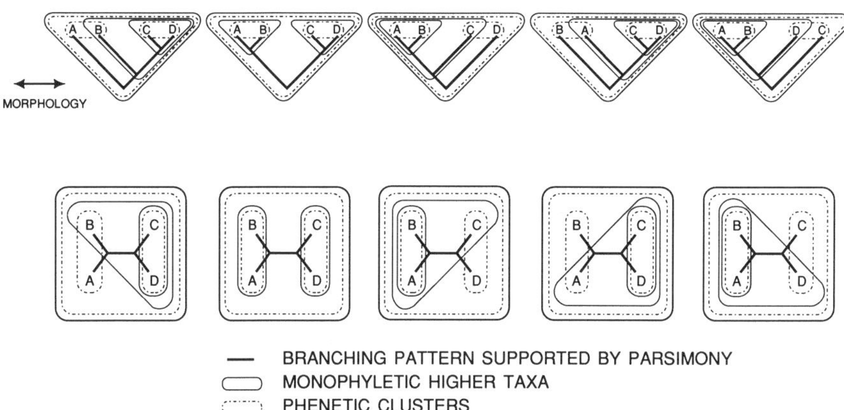

Figure 7. Cladogram selection for four taxa, with phenetic input. Unrooted networks, cladograms, and monophyletic groups are identical to those shown in Figure 6, but on cladograms, spacing on the horizontal axis indicates phenetic similarity. Phenetic groups are bounded by dashed lines, and in only one instance, the second set of diagrams from the left, are phenetic and monophyletic groups congruent. In this sense, assessing character state distributions for congruence and for phenetic resemblance resolves a unique hierarchy.

networks are identical to those of Figure 6. In Figure 7, however, phenetic resemblance is indicated, in the cladograms, by proximity in the horizontal direction, and phenetic groups are shown by the dot-dash boundaries. Focusing on the interaction of phenetic groups and the groups considered monophyletic by respective cladograms, an interesting pattern emerges: there is incongruence between higher taxa and phenetic groups for all hierarchies but one, the second from the left. In other words, assessing patterns of character distribution for both congruence and phenetic resemblance, and looking for the higher-order congruence between groups sorted by these criteria, *does* suffice to pick out a unique hierarchy from among all alternatives. It may not be the hierarchy picked using some other criterion, but it is unique. This solution is certainly in the spirit of pattern cladistics in that similarity is a relationship, not a process. It is a structural property of a system of entities, and as the basis for a convention, it may be used to group things without reference to a process assumption. It is also not necessary to assume any particular cause for observed similarities (though we may later need to defend our interest in similarity). There is indeed order here, and as Brady (1985) has argued, that order is an *explanandum* appropriately addressed by the *explanans* that Darwin offered us. But once again, the order does not emerge simply from perceiving congruence in character distributions; it is revealed here by the second-order congruence of patterns of character distribution and phenetic groups.

At this point, I need to make several disclaimers. First, I see no evidence that there was a *unique*, pre-Darwinian perception of the hierarchical organization of organisms. It would in fact be surprising if pre-Darwinian naturalists managed to achieve greater unanimity on this than have their successors. Second, I make no claim that pattern cladistics + phenetics, in the sense described above, was actually responsible (any more than I suspect of pattern cladistics + ontogeny) for whatever degree of consensus may be distilled from pre-Darwinian efforts to systematize life's diversity. Nor am I suggesting that pattern cladistics + phenetics provides an adequate perspective on the world as we see it today. I only note in passing, the ironic but fairly simple result that these two in combination accomplish what pattern cladistics by itself seems unable to do (without compromises of the sort discussed above). The point at which I believe the pattern cladistics + phenetics combination becomes problematic is when we finally have to justify the relevance of similarity. In the end, however, I believe even this is a tangential issue in comparison to another approach to discovering pattern in life's diversity.

PATTERN CLADISTICS + STRATIGRAPHY/TAPHONOMY

The alternative approach I wish to feature here is based on

interaction of patterns of character congruence and patterns of ordering of taxa in time. This is not entirely unrelated to notions such as Agassiz's parallelism between comparative morphology and paleontology, discussed by Nelson (1978), but the proposal offered here is fundamentally different in its actual content and manner of operation. The ordering of taxa in time is based on the extrinsic attribute of relative position within a composite stratigraphic sequence, and thus it carries information only when unit taxa differ in stratigraphic position. Fundamentally, and in the simple version presented here, unit taxa are taken to be samples derived from a restricted interval of time, so the operative stratigraphic information is (relative) position, not range. Use of stratigraphic data as a part of cladistic analysis distinguishes this approach from that taken by several recent demonstrations of the importance of fossils based on the (traditional morphological) character combinations they introduce (Gauthier et al., 1988; Donoghue et al., 1989; Novacek, 1992), although these demonstrations significantly enhance the background for the current proposal. Likewise, systematists' interest in stratigraphy may be attracted by the generally high correlation between stratigraphic and cladistic order (Norell and Novacek, 1992), and yet the current approach does not take this as a precondition.

This approach is illustrated in simple form in Figure 8, where vertical placement of terminal taxa on the cladogram represents their stratigraphic order. The solid boundaries mark higher taxa recognized by each cladogram, and for comparison against a constant reference, these are also displayed on the unrooted network below. The dashed boundaries indicate groups of taxa that appear *after* successive times chosen to correspond with just those stratigraphic positions at which we have recorded taxa, and these groups are transferred onto the networks too. The interesting pattern, wherever one prefers to trace it, is that on all but one pair of diagrams there is incongruence. Only on the central cladogram in Figure 8 (and on the unrooted network below it) is there a tidy nesting of putatively monophyletic groups and temporal groups. Things become even more interesting with larger numbers of taxa. The same conventions are applied in Figure 9, and once again, incongruence,

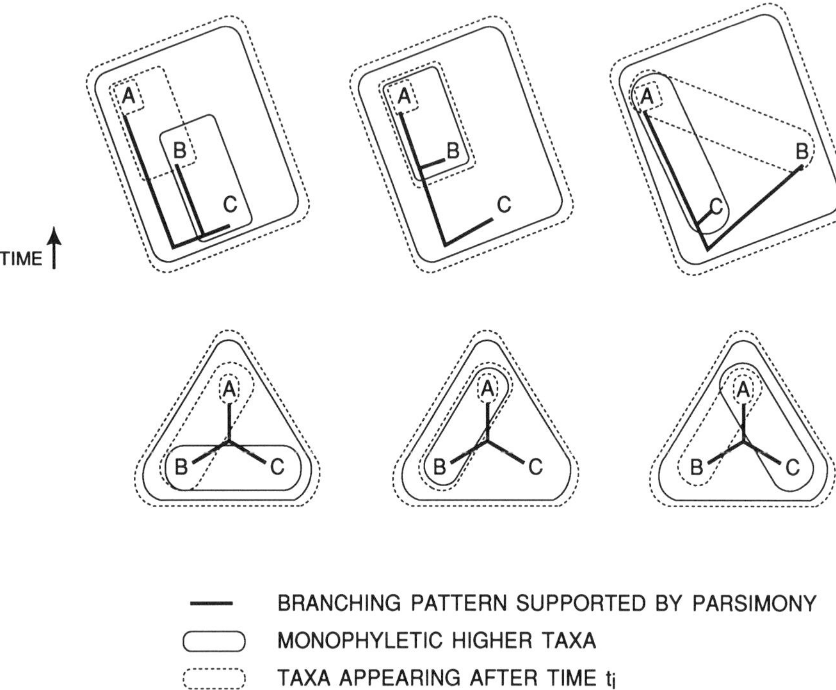

Figure 8. Cladogram selection for three taxa, with stratigraphic input. For the cladograms, temporal ordering of taxa is indicated by vertical position. Congruence (i.e., lack of intersecting boundaries) between putatively monophyletic groups and groups identified by the temporal convention applied here obtains for only one of the three alternative cladograms, shown in center position in the series of cladograms.

manifested by intersecting boundaries, rules out all hierarchies but one, again placed in center position. Furthermore, different incongruent hierarchies show different *degrees* of incongruence, with three cases of overlap on the leftmost pair of diagrams, one on the next pair to the right, one on the rightmost pair of diagrams, and three on the next pair to the left. The number of cases of overlap is a simple measure of incongruence comparable to the number of instances of homoplasy in evaluating character incongruence, and it reflects the degree of mismatch between the ordering of taxa in time and the cladistic ordering posited by a cladogram.

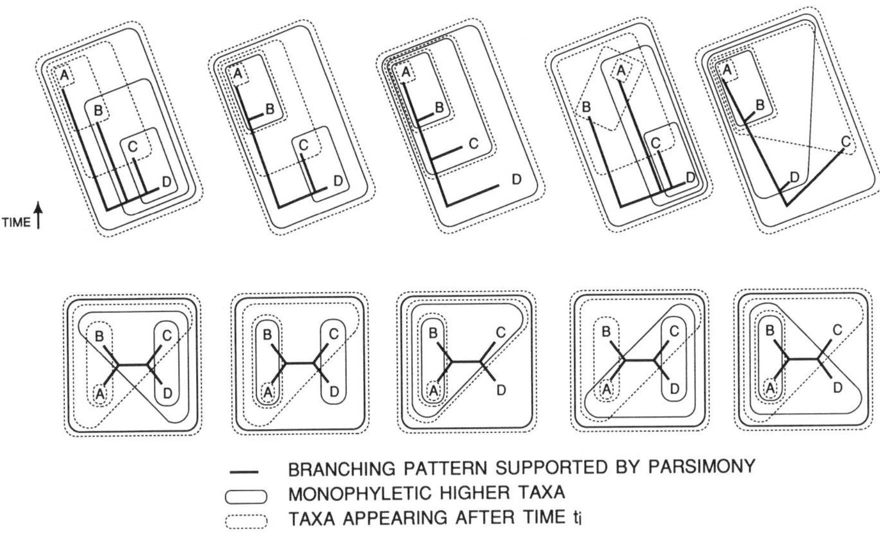

Figure 9. Cladogram selection for four taxa, with stratigraphic input. For the cladograms, temporal ordering of taxa is indicated by vertical position. Congruence (i.e., lack of intersecting boundaries) between putatively monophyletic groups and groups identified by the temporal convention applied here obtains for only one of the five alternative cladograms, shown in center position in the series of cladograms.

It may be asked, as indeed one reviewer did, whether we are here testing cladograms or the completeness of the fossil record. This is always an issue when we seek congruence between two bodies of data, be they different groups of morphologic characters, morphologic and molecular data, or character data and temporal ordering of occurrences. If we restrict ourselves to the current level of analysis, little more can be said except that hypotheses compatible with both sets of data will be preferred over ones incompatible with either, and the degree of incompatibility with each must be accounted for in any hypothesis choice. However, the completeness of the fossil record can be evaluated to some extent, independent of cladistic hypotheses, by turning outside systematics and biology altogether, to stratigraphy and taphonomy (Fortey and Jefferies, 1982). For instance, if the available stratigraphic sections do not, at a given level, represent facies necessary for preservation of some subset of the taxa in the analysis, then the record is at this point demonstrably incomplete, and absences of these taxa from this level cannot be counted against hypotheses. This property of having roots outside systematic biology is an important strength of stratigraphic data, discussed further below.

This joint evaluation of patterns of character congruence and patterns of ordering in time is what I have referred to previously as stratocladistics (Fisher, 1982, 1988, 1991), and what I have tried to illustrate here is a pattern cladist's door onto that subject. To paraphrase John Beatty's (1982, p. 31) quote of Colin Patterson, "in presenting it, I had no need to use...words like...paraphyly, speciation, dichotomy, ancestry, adaptation–don't need them. ... You don't need to know about evolution, or believe in it, to do [strato]cladistic analysis. All [strato]cladistics demands is that groups have characters [and times of occurrence], and that the groups are nonoverlapping." I admit to stating my position this way in part to highlight the irony that will be evident to all who are familiar with Colin Patterson's views on the role of fossils in our perception of life's order (Patterson, 1981), but this implies no frivolity in my argument that patterns such as those in Figures 8 and 9 are capable of resolving hierarchy within organic diversity.

To illustrate the form of this argument, I have used simple examples, leading to a completely unproblematic resolution. More complex examples (e.g., with multiple taxa at some or all stratigraphic positions) would display more complex behavior (e.g., requiring a temporal criterion that includes reference to characters and allows multiple groups of taxa occurring after time t_i; this is in part why I propose use of form + time, but never time alone, to group taxa). Just as it is unnecessary for all character patterns to be perfectly congruent with each other, it is unnecessary for stratigraphic patterns to be perfectly congruent with character patterns. We simply look for the pattern that dominates and try to minimize the number of instances of incongruence in our interpretation of the total data set. Single stratigraphic incongruities do not necessarily rule out cladistic groupings; they only provide a measure of congruence to be used in the comparative evaluation of all viable hypotheses. For these examples, I have assumed that morphologic (intrinsic character) congruence was so overwhelming that it restricted attention entirely to cladograms consistent with a single unrooted network, but in actual practice, stratigraphic incongruities may offset some or all of this difference in degree of morphologic congruence. If stratigraphic patterns are strong enough, they may overturn cladograms (and unrooted networks) supported by morphology alone (Fisher, 1991, 1992). In any event, stratocladistics does not force a literal reading of the stratigraphic record; it simply asks that we treat incongruities between the stratigraphic record and taxonomic groups in the same way we would treat incongruities within character data–as elements of our interpretation that can be accommodated and set aside if necessary, but not, in general, without introducing an ad hoc element into the analysis. The search for maximal congruence is equivalent to minimizing reliance on such ad hoc elements (Farris, 1983).

It is important to note, by way of distinction from other procedures and arguments, that searching for second-order congruence between taxa (identified by first-order congruence of character distributions) and temporal groups as suggested here, is *not* equivalent to using stratigraphic data to polarize characters. The procedure discussed here embodies no claim about character polarity per se (Nelson, 1978), and the count of instances of incongruence, as a measure of the merit of competing hypotheses of grouping, behaves differently from a count of instances in which character polarities disagree with order of stratigraphic occurrence. I should also clarify that counting instances of incongruence as done here is not precisely equivalent to what I have defined previously as "stratigraphic parsimony debt" (Fisher, 1991, 1992). It is, however, a component of stratigraphic parsimony debt –more precisely, that component that addresses congruence between cladistic ordering and order of occurrence. The component of stratigraphic parsimony debt that is excluded from consideration here (not because it is unimportant or uninteresting, but rather only because it is a separate issue) is that which reflects the relation between temporal ordering and the possibility of ancestor-descendant relationships between taxa. Essentially, I have chosen here to focus on the way stratigraphic patterns might help to choose among alternative cladograms, and I leave for separate treatment the simpler problem of how they might help to choose among alternative phylogenetic trees associated with a given cladogram(s). In many other accounts, this latter problem is the only context in which stratigraphic data are allowed a role, if they are entertained at all (e.g., Eldredge and Cracraft, 1980). The stratigraphic record certainly provides our best evidence of the "temporal boundedness of individual taxa" (Eldredge and Novacek, 1985), but despite early pronouncements to the contrary (e.g., Schaeffer et al., 1972), it can also participate in the recognition of those taxa.

BIOGEOGRAPHIC DATA

Systematists are interested in patterns in space (e.g., Nelson and Platnick, 1981; Myers and Giller, 1988) as well as time, and stratigraphy, as treated by geologists, involves both. Therefore, although biogeographic distributions are not the main focus of this discussion, I will digress to consider how they relate to the patterns discerned thus far. This treatment passes by many issues targeted in recent contributions to biogeography but is intended only as brief reconnaissance of an area requiring more intensive study. Stratocladistics, even with extension to the spatial domain, is not equivalent to Croizat's (1958) panbiogeography, but it shares with the latter the idea that space, time, and form combine to reveal informative patterns in nature.

Figure 10 presents a simple case involving three taxa illustrating a vicariance event. This diagram is organized much like Figure 8, with time as the vertical dimension of the cladograms, except that here, A and B occur at essentially the same time, and C occurs earlier. Groups are circumscribed by boundaries as before, but the different dimensionality of space and time has interesting consequences for the set relations of different groups defined by a single criterion, operating on a single modality (space, time, or form; remember that more complex problems will require compound criteria, e.g., time + form, space + form). Since time is unidimensional, groups of taxa "appearing after time t_i" for a later t_i, are automatically nested within such groups defined for an earlier t_i. Hierarchical structure is inevitable, and as discussed above, either will or will not be congruent with the hierarchical structure of a given cladogram. With geography, groups defined by their distributions may, but need not, be hierarchically nested. If we are interested in any hierarchical patterns that may exist, it is necessary to record them in an unambiguous way. The geographic groupings in Figure 10, treated as Venn diagrams, must therefore reflect, if only qualitatively, the set relations of taxon ranges. In this case, A and B are distributed in nonoverlapping areas, each of which is a proper subset of the area inhabited by C, and this is shown on the diagrams, along with putatively monophyletic and temporal groups. Only one cladogram of the three shows congruence of all these patterns. Not surprisingly, it is the one corresponding to the vicariant model set up in the first place. It seems at first glance that biogeography has contributed nicely to the resolution of pattern, but it is important to note that the geographical data provided none of the incongruence that caused the rejection of two of the alternatives. The distributions of A, B, and C are compatible with all of the cladograms. Thus, although vicariance involves regions that relate in an inherently hierarchic way, in this instance it does not provide observations that allow us to discriminate between alternatives for ordering taxa.

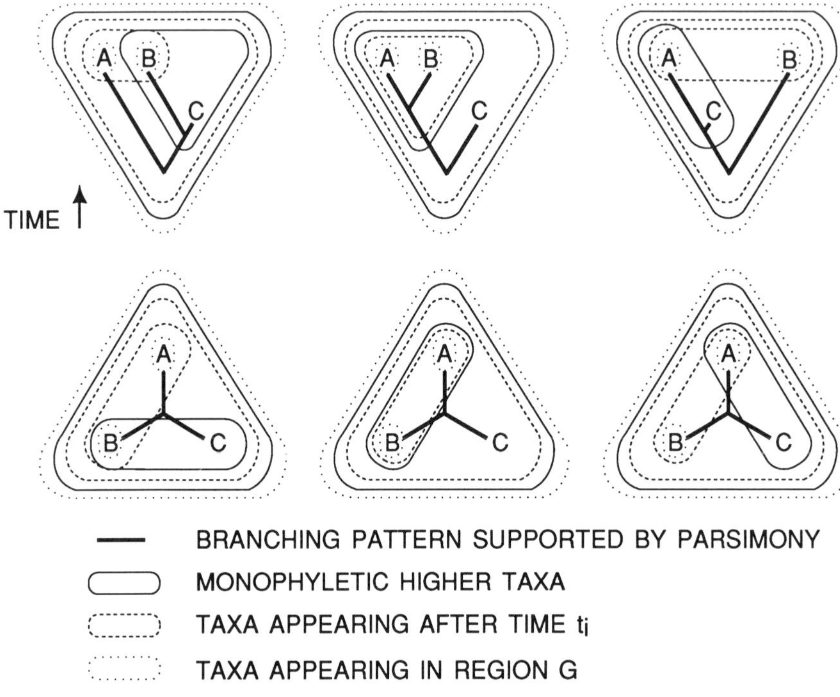

Figure 10. Cladogram selection for three taxa, with stratigraphic and biogeographic input. The example treated is a simple instance of vicariance. For the cladograms, temporal ordering of taxa is indicated by vertical position. Congruence (i.e., lack of intersecting boundaries) between putatively monophyletic groups, groups identified by the temporal convention applied here, and groups whose distribution is restricted to a given region (with subset relations on the diagrams corresponding to spatial structure) obtains for only one of the three alternative cladograms, shown in center position in the series of cladograms.

For the next step, I wish to set temporal data temporarily aside and deal only with taxonomic groups and their biogeographic ranges.

The question of interest is whether biogeographic histories other than vicariance can be resolved in this way. I will consider first a history of extirpation, or range contraction. I realize that this approach involves starting with a process and asking what pattern might be associated with it, but my goal is to explore a variety of patterns and determine what implications they may bear. In progressive range contraction we have a process that produces areas that are automatically nested. The groups of taxa associated with these areas either will or will not be congruent with the monophyletic groups of a given cladogram, and in this case, the groups based on area do refute all cladograms but the "correct" one around which this scenario was constructed (Fig. 11).

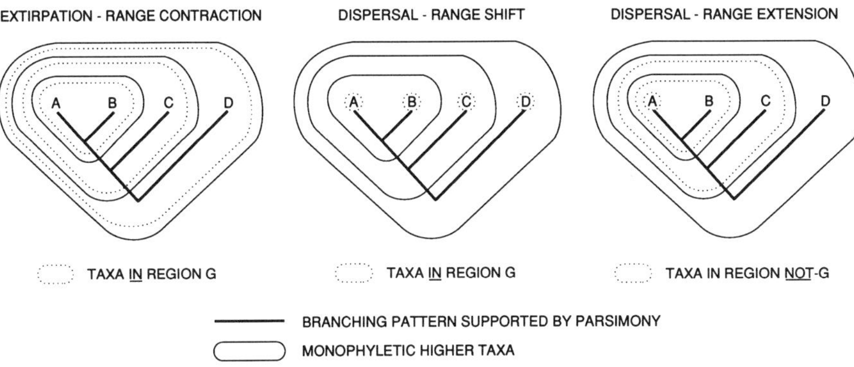

Figure 11. Congruent patterns between higher taxa and geographically defined groups of taxa, for three different biogeographic scenarios.

Another model, dispersal involving progressive range shift, produces areas that are not hierarchically related. As such, these areas do not generally define hierarchically organized groups of taxa other than the trivial ones of individual unit taxa within the overall group. This is uninformative in the sense that it does not refute groupings of taxa proposed by competing cladograms retained by the criterion of first-order character congruence (Fig. 11).

Dispersal involving range extension again produces areas that are hierarchically organized, but in this case, areas (as before, the areas inhabited by specified unit taxa or groups of unit taxa) are nested *backward* in time (earlier within later). Groups defined in this way do succeed in picking, by second-order congruence, one of the competing cladograms retained by the criterion of first-order character congruence, but it is the "wrong" one–i.e., one not equivalent to the scenario on which this example is based. However, an alternative approach to identifying areas associated with particular taxa does work. First, identify the order in which taxon areas nest; then group together not taxa that live in particular areas, but rather, taxa that inhabit areas outside, or beyond, the area inhabited by the more geographically restricted group. This is indicatcd in Figure 11 as "taxa in region not-G." This hierarchical system also suffices to pick out a single cladogram (of the five competitors identified, as in Figure 6, by first-order congruence), and this time it is the "right" one.

Figure 11 thus shows congruent patterns between higher taxa and geographically defined groups of taxa, but in order to know which geographic grouping to construct, we had to have an idea of what historical pattern we were dealing with. The situation becomes much more tractable, however, if we have temporal data (Fig. 12). Extirpation and the cladogram for taxa showing such a pattern are identified by congruence of putative higher taxa, taxa appearing after t_i, and taxa in particular regions. Cladogram choice in a range shift context is based, as above, on congruence of putative monophyletic higher taxa and taxa appearing after t_i; geographic data do not assist in this choice, but congruence between taxa inhabiting particular areas and taxa appearing between t_i and t_{i+n} defines the scale and rate of the range shift. Range extension is marked by congruence of putative higher taxa, taxa appearing after t_i, and taxa in region not-G. How would we know what type of biogeographic data to use (taxa in region G or taxa in region not-G)? In general, there will be some scale of problem-definition at

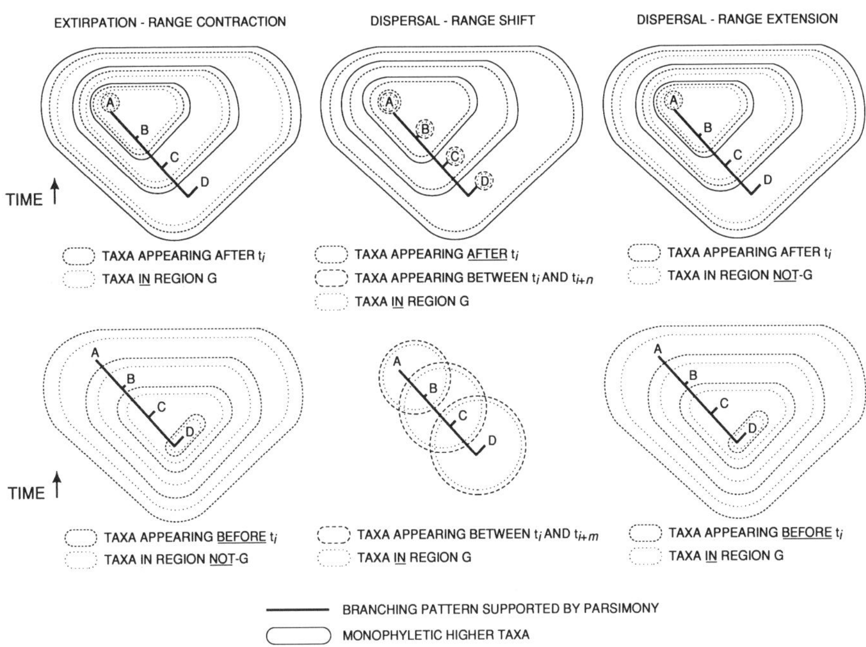

Figure 12. Congruent patterns between higher taxa, geographically defined groups of taxa, and temporally defined groups of taxa, for three different biogeographic scenarios.

which only one of these will be congruent with groups of taxa appearing after t_i. Alternatively, we can look at groups of taxa appearing before t_i, and this switches the association between process (range contraction or range extension) and the type of geographic pattern utilized (taxa in region G or taxa in region not-G). Following this route yields congruent temporal and geographic groupings, but ones that consist of paraphyletic, not monophyletic, higher taxa. Likewise, range shift, evaluated at a different scale (relative to taxa) from the case above, yields paraphyletic groups discoverable by congruence of biogeographic and temporal (taxa appearing between t_i and t_{i+m}) patterns. Although

I do not intend to press this point here, it raises the possibility that paraphyletic groups are not bereft of reality after all.

To conclude this section, it should be noted that evaluation of both biogeographic patterns and temporal patterns involves assessing where taxa are *not* located in space-time as well as where they are. Is this commerce with negative evidence crippling (e.g., Rosen's comments on the "ambiguity of absence"; 1988, p. 448)? Is it even negative evidence in the usual sense of the word? Undoubtedly, the quality and resolution of both the biogeographic and the temporal record are variable, but in many instances (and increasingly so as we move to larger scales of analysis), records of distributions are not simply tallies of encounters with isolated specimens. Rather, they are often records of many specimens of certain taxa, occurring in certain environments, with no specimens of certain other taxa. Furthermore, within such assemblages we can frequently distinguish what are known as taphonomic controls (Jablonski and Bottjer, 1991), taxa with similar preservation/recovery probabilities to those whose absence is of potential significance. Where absences are not significant on stratigraphic and taphonomic grounds, they should not be counted against a hypothesis (Fisher, 1991, 1992). These sorts of comparative observations are the basis for the biogeographic and temporal patterns I have discussed here. I do not mean to suggest that their analysis does not involve uncertainty, but the uncertainties faced are qualitatively similar for both biogeographic and temporal data, from both the fossil record and the present.

PATTERN-PROCESS INTERACTION

The type of analysis illustrated here can be characterized fairly, I believe, as a "discovery procedure" (Nelson, 1989) focused on discerning pattern in life's diversity; its application resolves one (or in less clear-cut cases, more than one) pattern among many initial possibilities. Relative stratigraphic position is admittedly an extrinsic attribute (Eldredge and Cracraft, 1980), but it is no less subject to objective documentation on this account. Use of stratigraphic data along with intrinsic character data also represents a move in the direction of treating "total evidence" (Kluge, 1989), although the call for this has not been a pattern cladist preoccupation. Why should a pattern cladist be interested in temporal order? –partly because it works, as demonstrated to some extent in the examples presented here, often to corroborate, sometimes to refine, and in certain circumstances to refute, the groupings suggested by character distributions. Second, stratocladistics as conducted here behaves in a formal sense much like pattern cladistics. Like grouping by congruence, or by presence of synapomorphies, temporal order provides a basis for grouping taxa that does not explicitly refer to process (though I will argue below that all such procedures do so implicitly). It constitutes a structural relationship among the attributes of taxa that, like a series of footprints, can be described at some level without presuming knowledge of its cause.

Despite this compliance with the pattern cladist characterization of systematic analysis, the apparent independence of data and hypothesis deserves closer scrutiny. I have tried to present a pattern cladist perspective, but in the end, I cannot agree that our world view ever emerges without input from notions of process. In part, these notions enter when we try to defend the relevance of the criteria we bring to bear in the recognition of pattern. If pattern cladists fail to take this step, they commit themselves to an arbitrary scheme and risk ignoring other relevant observations. Such interaction between pattern data and process hypotheses should not really be a problem, even for pattern cladists, because hypotheses do not *have* to be built from the ground up. As hypotheses, they are free-floating ideas that must make their own peace with patterns in the data, but need not be deduced (or even induced) directly from such patterns. Which patterns are relevant is determined by the hypothesis, not by a priori reflection.

If pattern cladists do attempt to relate their criteria to process, I argued above that whatever process hypotheses are actually entertained need to "cover the same ground" that phylogenetic (lineage-level) and evolutionary (within-lineage) processes cover in evolutionary models. Whether grouping is accomplished by identifying features present in organisms, by setting up hierarchical relations that mirror ontogeny, or by using outgroups identified in prior analyses based on the same or different methods, notions of process influence perception of life's pattern. This influence is felt in decisions related to character analysis and decisions implied by our treatment of characters as independent (or *not*, as the case may be) and of equal (or *un*equal) evidential value (Sober, 1988). Even when we do not "say the words," our treatment of characters implies substantive positions on issues of how different character states are related (in an abstract, not a phylogenetic, sense), and that has very much to do with whatever process is responsible for the diversity of states we observe.

I accept this same burden when it comes to stratigraphic data. The relevance of temporal ordering is a separate matter from its effectiveness in sorting among alternative hierarchies and certainly derives from hypotheses involving phylogeny. Descent, branching, and transformation are time-dependent processes and imply an ordering of taxa in time. Furthermore, competing hypotheses of relationship imply different temporal orderings of taxa and thus different expected temporal patterns of occurrence. The observed temporal pattern of occurrence of taxa may thus differentially support competing hypotheses of relationship as well as provide constraints on the absolute timing of phylogenetic events. Roughly speaking, this takes us from the level of lineages to the taxonomic hierarchy (Process_1 or Pattern_2 to $\text{Pattern}_{1.5}$ in Fig. 3). It deals with process to the extent of suggesting that the taxa whose attributes we code and whose relationships we seek to resolve are linked by a pattern of incompletely recorded lineages within which attributes have changed ("descent with modification"). However, pattern cladists implicitly, and phylogenetic systematists more or less explicitly, also deal with the span of levels of analysis from within-lineage to lineages (Process_2 to Pattern_2 or Process_1 in Fig. 3). This deals with how attributes change within lineages and what processes determine or link the attributes we record. Stratocladists must cover this same span, but at this point there is a fundamental difference. There are no mysteries about why descendants postdate their ancestors comparable to the questions still open about why descendants sometimes (but to varying degrees) differ in form from their ancestors. The temporal relations of segments of a branching system of lineages are understood as a function of the directionality of time. This, of course, is not all that affects "the attributes we record"–treatment of stratigraphic data also depends on our understanding of what controls the stratigraphic occurrence of fossils. Here too, though, the processes affecting stratigraphic patterns are actually fairly well understood. They are processes of surficial weathering, transport, burial, and diagenesis–in short, stratigraphy and taphonomy. These are understood in part from first principles, but can also be addressed empirically, within contexts of particular interest, by evaluation of taphonomic controls (Jablonski and Bottjer, 1991). As a result, geologists *now* understand more about the factors that control stratigraphic distribution than biologists are likely to understand anytime soon about the factors that control the character states of organisms. Stratocladistics thus depends on an understanding of process no less than pattern cladistics or phylogenetic systematics, but to the extent that it draws its resolution from stratigraphic data, it depends on an understanding of *different* processes. Use of extrinsic attributes for phylogenetic analysis has been criticized because they do not participate in a "normal" style of descent with modification (Eldredge and Cracraft, 1980), but epistemologically, this is a liberating rather than a damning quality.

In contrast to the pattern cladist view, I do not think we can escape notions of process, but stratocladistics at least draws much of its resolution through process assumptions from *outside* evolutionary theory. As such, it has within its grasp a degree of independence from the processes whose workings it seeks to reveal, that other approaches to systematics have claimed, but none has matched.

Acknowledgments

I thank Olivier Rieppel for inviting this contribution, and the other symposium participants for the stimulus provided to these thoughts by their own work. I also greatly appreciate comments from B.

Chernoff, J. J. Flynn, P. D. Gingerich, O. Rieppel, and G. R. Smith during the development of these ideas. M. J. Benton, M. J. Foote, O. Rieppel, G. R. Smith, M. Wilkinson, and especially C. Patterson and an anonymous reviewer criticized a draft of the manuscript and helped clarify arguments greatly, but cannot be held responsible for the heterodoxies that remain.

References

Beatty, J. 1982. Classes and cladists. *Systematic Zoology*, 31(1): 25-34.

Brady, R. H. 1985. On the independence of systematics. *Cladistics*, 1(2): 113-126.

Croizat, L. 1958. *Panbiogeography*. Caracas: L. Croizat.

Darwin, C. 1859. *On the Origin of Species*. London: Murray.

de Queiroz, K. 1985. The ontogenetic method for determining character polarity and its relevance to phylogenetic systematics. *Systematic Zoology*, 34: 280-299.

de Queiroz, K., and M. J. Donoghue. 1990. Phylogenetic systematics or Nelson's version of cladistics? *Cladistics*, 6: 61-75.

Donoghue, M. J., J. A. Doyle, J. Gauthier, A. G. Kluge and T. Rowe. 1989. The importance of fossils in phylogeny reconstruction. *Annual Review of Ecology and Systematics*, 20: 431-460.

Eldredge, N., and J. Cracraft. 1980. *Phylogenetic Patterns and the Evolutionary Process*. New York: Columbia University Press.

Eldredge, N., and M. J. Novacek. 1985. Systematics and paleobiology. *Paleobiology*, 11(1): 65-74.

Farris, J. S. 1983. The logical basis of phylogenetic analysis. In N. I. Platnick and V. A. Funk (Eds.), *Advances in Cladistics*, Vol. 2, 7-36. New York: Columbia University Press.

Fisher, D. C. 1982. Phylogenetic and macroevolutionary patterns within the Xiphosurida. *Proceedings of the North American Paleontological Convention III*, 1: 175-180.

Fisher, D. C. 1985. Evolutionary morphology: beyond the analogous, the anecdotal, and the ad hoc. *Paleobiology*, 11(1): 120-138.

Fisher, D. C. 1988. Stratocladistics: integrating stratigraphic and morphologic data in phylogenetic inference. *Geological Society of America, Abstracts with Programs*, 20(7): A186.

Fisher, D. C. 1991. Phylogenetic analysis and its application in evolutionary paleobiology. In N. L. Gilinsky and P. W. Signor (Eds.), *Analytical Paleobiology*, 103-122. Short Courses in Paleontology, No. 4, Paleontological Society.

Fisher, D. C. 1992. Stratigraphic parsimony. In W. P. Maddison and D. R. Maddison, *MacClade: Analysis of Phylogeny and Character Evolution, Version 3*, 124-129. Sunderland, MA: Sinauer Associates Inc. 398 pp.

Fortey, R. A., and R. P. S. Jefferies. 1982. Fossils and phylogeny–a compromise approach. In K. A. Joysey and A. E. Friday (Eds.), *Problems of Phylogenetic Reconstruction*, 197-234. New York: Academic Press.

Gauthier, J., A. G. Kluge and T. Rowe. 1988. Amniote phylogeny and the importance of fossils. *Cladistics*, 4: 105-209.

Gingerich, P. D. 1979. Stratophenetic approach to phylogeny reconstruction in vertebrate paleontology. In J. Cracraft and N. Eldredge (Eds.), *Phylogenetic Analysis and Paleontology*, 41-77. New York: Columbia University Press.

Hennig, W. 1966. *Phylogenetic Systematics*. Urbana: University of Illinois Press.

Hull, D. L. 1967. Certainty and circularity in evolutionary taxonomy. *Evolution*, 21: 174-189.

Jablonski, D., and D. J. Bottjer. 1991. Environmental patterns in the origins of higher taxa: the post-Paleozoic fossil record. *Science*, 252: 1831-1833.

Kluge, A. G. 1985. Ontogeny and phylogenetic systematics. *Cladistics*, 1: 13-27.

Kluge, A. G. 1988. The characterization of ontogeny. In C. J. Humphries (Ed.), *Ontogeny and Systematics*, 57-81. London: British Museum (Natural History).

Kluge, A. G. 1989. A concern for evidence and a phylogenetic hypothesis of relationships among *Epicrates* (Boidae, Serpentes). *Systematic Zoology*, 38: 7-25.

Kluge, A. G. 1991. Boine snake phylogeny and research cycles. *Miscellaneous Publications, Museum of Zoology, University of Michigan*, 178: 1-58.

Lundberg, J. G. 1972. Wagner networks and ancestors. *Systematic Zoology*, 21: 398-413.

Mabee, P. M. 1989. An empirical rejection of the ontogenetic polarity criterion. *Cladistics*, 5: 409-416.

Meacham, C. A. 1984. The role of hypothesized direction of characters in the estimation of evolutionary history. *Taxon*, 33: 26-38.

Myers, A. A., and P. S. Giller (Eds.). 1988. *Analytical Biogeography: an Integrated Approach to the Study of Animal and Plant Distributions*. London: Chapman and Hall.

Nelson, G. 1978. Ontogeny, phylogeny, paleontology, and the biogenetic law. *Systematic Zoology*, 27(3): 324-345.

Nelson, G. 1989. Cladistics and evolutionary models. *Cladistics* 5: 275-289.

Nelson, G., and N. Platnick. 1981. *Systematics and Biogeography: Cladistics and Vicariance*. New York: Columbia University Press.

Norell, M. A., and M. J. Novacek. 1992. The fossil record and evolution: comparing cladistic and paleontological evidence for vertebrate history. *Science*, 255: 1690-1693.

Novacek, M. J. 1992. Fossils, topologies, missing data, and the higher level phylogeny of eutherian mammals. *Systematic Biology*, 41: 58-73.

Patterson, C. 1981. Significance of fossils in determining evolutionary relationships. *Annual Review of Ecology and Systematics*, 12: 195-223.

Patterson, C. 1982. Morphological characters and homology. In K. A. Joysey and A. E. Friday (Eds.), *Problems of Phylogenetic Reconstruction*, 21-74. London: Academic Press.

Rieppel, O. 1985. Ontogeny and the hierarchy of types. *Cladistics*, 1(3): 234-246.

Rieppel, O. 1990. Ontogeny–a way forward for systematics, a way backward for phylogeny. *Biological Journal of the Linnean Society*, 39: 177-191.

Rosen, B. R. 1988. From fossils to earth history: applied historical biogeography. In A. A. Myers and P. S. Giller (Eds.), *Analytical Biogeography: An Integrated Approach to the Study of Animal and Plant Distributions*, 437-481. London: Chapman and Hall.

Schaeffer, B., M. Hecht and N. Eldredge. 1972. Phylogeny and paleontology. *Evolutionary Biology*, 6: 31-46.

Sober, E. 1988. *Reconstructing the Past: Parsimony, Evolution, and Inference*. Cambridge, MA: Massachusetts Institute of Technology Press.

Wiley, E. O., D. Siegel-Causey, D. R. Brooks and V. A. Funk. 1991. *The Compleat Cladist: a Primer of Phylogenetic Procedures*. University of Kansas Museum of Natural History, Special Publication No. 19, Lawrence, KS.

7

The Use of Unconventional Morphological Characters in Analysis of Systematic Patterns and Evolutionary Processes

Marvalee H. Wake

Department of Integrative Biology and
Museum of Vertebrate Zoology
University of California, Berkeley, California 94720

Abstract. Many functional morphologists now hypothesize patterns of evolutionary change in taxa or functional morphological complexes by mapping these complexes on pre-existing cladograms. The use of cladograms as tools for such analyses requires that functional morphologists understand cladistic methodology. The linkage of character transformation hypotheses to phylogenetic patterns will result in a better understanding of the evolution of functional morphology.

Morphologists have not yet completely exploited the utility of morphology for analyzing pattern and process. The *biology* of certain morphological features should be better understood in order to fully evaluate their potential as characters for phylogenetic analysis. Examples are presented that raise new concerns about old questions: how we identify discrete characters within a functional complex and how these characters and their integration into a functional complex evolve. There is new interest in the origin of characters and their homologies, and the question of homology in phylogenetic analysis. Finally, more attention should be paid to the arguments of structuralists who claim that the assumption of phylogenetic systematics (that morphological relationships should be congruent with genealogical relationships) is erroneous. They see a fundamental disparity between genealogy and the taxonomy of forms, and infer that genetic evolution is decoupled from morphological evolution. This may explain certain incongruencies between phylogenetic patterns when one is based on molecular data while the other is based on morphological data.

Copyright © 1994 by Academic Press, Inc.
All rights of reproduction in any form reserved.

Introduction

THE EVOLUTION OF COMPLEX FUNCTIONAL SYSTEMS (e.g., morphological, ecological or behavioral) is best studied with reference to a phylogenetic framework (cladogram). This raises the question as to how the study of transformation of functional morphological complexes relates to pattern and process. On one hand, morphology is used as database for pattern reconstruction, but on the other hand, phylogenetic patterns constrain our theories of morphological transformation. To avoid circular reasoning, the pattern used to constrain hypotheses of evolutionary change of morphological systems must be based on independent evidence. This requires that functional morphologists using cladograms generated by different authors understand cladistic methodology, as well as the characters used to generate these cladograms.

Further, I contend that morphologists have not yet completely explored the use of morphology for analyzing pattern and process. Vertebrate morphologists have relied on external features, and bones and muscles; invertebrate morphologists have looked carefully at external features, such as the genitalia. But neuroanatomical patterns, sperm morphology, and other features of soft anatomy that are less well understood in many ways, are just beginning to be assessed carefully (there are some exceptions to these generalizations). As a consequence of exploring "new" or "unconventional" morphological features and especially their integration into interactive complexes, there arise new concerns about some old questions: how to identify discrete characters and how to analyze them in an evolutionary context. There is renewed interest in the origin of characters and their homologies, and the question of homology and its role in phylogenetic analysis. Also, the perspective of the so-called "structuralists" must be addressed. Structuralists define taxonomic groups through logical relations of form without resorting to evolutionary process theories (Goodwin, 1984) thereby implicitly rejecting the assumption of phylogenetic systematics that morphological relationships, or relationships of form, should be congruent with genealogical relationships. These workers see a fundamental disparity between genealogy (based on molecular data) and the taxonomy of forms (morphological data), which implies to me that molecular and morphological evolution are decoupled.

The definition and recognition of homology is another concern in the analysis of characters. There has long been a debate about the nature of homology (recently discussed by Rieppel, 1988; Roth, 1984, 1988; Van Valen, 1982; Wagner, 1989; Wiley, 1981). Nelson (1970) considered homology an abstraction, but later referred to it as a historical relation (Nelson, 1989). Wagner (1989) provides a useful assessment of biological homology and the origin of morphological characters relevant to mechanistic concepts of evolutionary biology. He contends that one must consider the evolution of morphological characters at two levels: that of character modification within a framework of developmental constraints, and that of the level of the evolutionary modifications of constraints. The origin of new characters is associated with the origin of a developmental network that includes the subcharacters of the new trait. This concept is especially important as we consider the inclusion of morphological features in functionally complex units and try both to sort independent characters within such units and to use the units as characters. Understanding of homology will be even more important as we use molecular, behavioral, and ecological traits in phylogenetic analysis. New data, especially more biological information (developmental, ecological, etc.) about characters, can alter understanding of homology, and hence congruence of characters in a phylogenetic context. We only approach a real understanding of homology, but we have a better appreciation of its significance as well as its difficulty in identification. My thesis is that morphologists and systematists must better understand the *biology* of features in order to use them methodologically as characters (homologies) for phylogenetic analysis.

Felsenstein (1988) objects to what he calls the "discretization" of the data into two or a few states. He rues the loss of information in this method of character description. He views parsimony methods as making no assumptions about interdependencies of characters, thus reassuring the worker who knows that the measurements that underlie the discrete states of characters are indeed interdependent. Felsenstein recommends generation of a statistical framework, in which it is important to know of genetic, environmental, and selection pressure interdependencies. However, he acknowledges that the first two are rarely available to systematists, and the last almost never. At this stage, we must simply ask whether discrete coding is the best approach to characters, and how to deal with the possibility of interdependence in change of different characters. There are not yet clear answers to these questions.

PHYLOGENETIC ANALYSIS

Morphologists have not fully explored morphology for useful phylogenetic characters. We are beginning to see the use of new morphological (and other) data, in phylogenetic analysis. A few examples from the literature and from my research illustrate both the utility of these data and the problems they introduce to phylogenetic analysis. Morphologists, and other biologists, recently have employed phylogenetic analyses in two ways: (1) mapping of their data on cladograms generated from other databases in order to assess the congruence of their data with phylogenetic patterns, thereby to analyze the evolutionary pattern of the acquisition and maintenance of the traits; (2) to use those data to generate new cladograms to resolve questions of relationships, either by adding their new data to existing data sets, or by generating cladograms based solely on the independent data set, often followed by comparison to other cladograms. The former approach has been used by several morphologists recently to examine patterns of acquisition and evolution of traits, including functional, in recent and fossil forms (Gans and Northcutt, 1983; Northcutt and Gans, 1983; Wake and Larson, 1987; Hickman, 1988; Lauder and Liem, 1989; Liem, 1989; Lauder, 1990; Wake, 1991, to cite but a few); the approach recently has been championed by Brooks and McLennan (1991) as useful for analysis of evolution of ecological and behavioral features as well. The second approach is exemplified by several systematists, including Kluge (1989), who believes that all available data, morphological, molecular, and other, should be combined into a single data set for phylogenetic analysis. Other workers (see Cases 1, 2, and 3 below for examples) find it instructive to analyze the data sets separately, then perhaps together, to reveal incongruencies in the analysis. Many (Cannatella, pers. commun.) believe that adding new data to old data sets to produce new cladograms is more preferable than making new cladograms with the new data to compare to old cladograms. All of these systematists believe that the points of noncongruence indicate areas that need further study. It is in these activities that the critical choice of cladograms, understanding of methodologies and databases, and careful assessment of new traits, are required. A few examples illustrate both the utility and the pitfalls of phylogenetic approaches to character analysis.

Case 1: Bat Neuroanatomy

A debate is taking place in the pages of *Systematic Zoology* and other journals about the monophyly versus convergent evolution of bats (Fig. 1). Pettigrew (1991a, 1991b) and Pettigrew et al. (1989) base much of their argument for independent origins of megachiropterans (and primates) and microchiropterans on new (i.e., "unconventional") data from the nervous system, especially visual pathways, the corticospinal motor system, the lateral geniculate nucleus, and the hippocampus (finding 24 characters in all) (Fig. 2). Pettigrew concludes that neuroanatomical features, and a few others, support diphyly, but features of the wings are convergent, with emphasis on the idea that features of functional and adaptive import (such as wings) are subject to higher levels of homoplasy than other characters, and should not carry as much weight as "functionally obscure" characters. Baker et al. (1991) criticize Pettigrew and his colleagues for over-emphasizing neural structures (perhaps the *a posteriori* weighting sought by Felsenstein, but I doubt it), and specifying differences rather than derived similarities in wing structure, as well as their lack of alternatives for implicit weighting of conflicting characters, and the scope of their comparisons and outgroup assumptions. They argue that "functionally obscure" characters of the

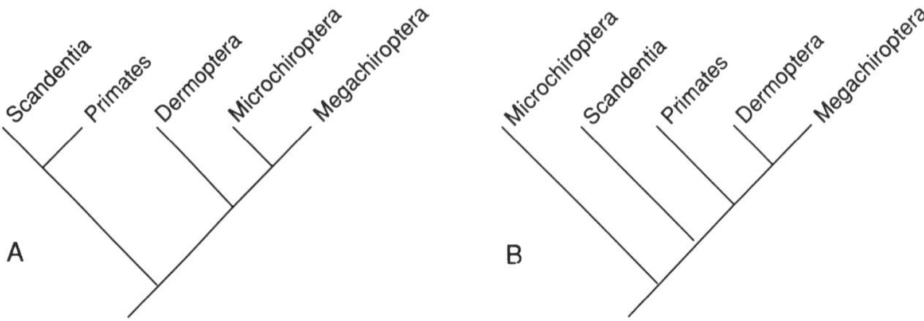

Figure 1. Competing hypotheses of phylogenetic relationships of bats and other mammals (after Baker et al., 1991). (A) "Classical" hypothesis; (B) "Alternative" hypothesis. See text for discussion.

neural system are not necessarily superior to other characters; in fact, we simply may not yet really understand the characters and their biology. Further, Baker et al. (1991) show that 12 of the 24 neural characters relate to the dorsal lateral geniculate nucleus; laminar differentiation of the nucleus is considered a derived state. Of the remaining 11 characters of the geniculate nucleus, eight depend on the presence of laminar differentiation. Pettigrew's character scoring of primitive and derived states does not treat several conditions, and obscures some of his own data, according to Baker et al. (1991). They contend that many reversals are needed to support Pettigrew's analysis, hence the emphasis on homoplasy and non-homology. Character scoring is crucial in analysis of character evolution and support for nodes of the cladogram. Baker et al. (1991), based on a large data set including a number of systems, conclude that the case for bat diphyly has not been established. Both Pettigrew and Baker et al. analyze molecular data as well, and find support for their respective positions but agree that the data are inconclusive. Both parties want more comprehensive analyses of morphology and of molecular structures in a wider range of taxa, both want exploration of new character systems, and Simmons et al. (1991) request more explicit information about the characters. Full presentations of data, methodologies, and analyses are crucial to resolution of this problem. The characters may be nonindependent, with a causative relationship, or they may be independent.

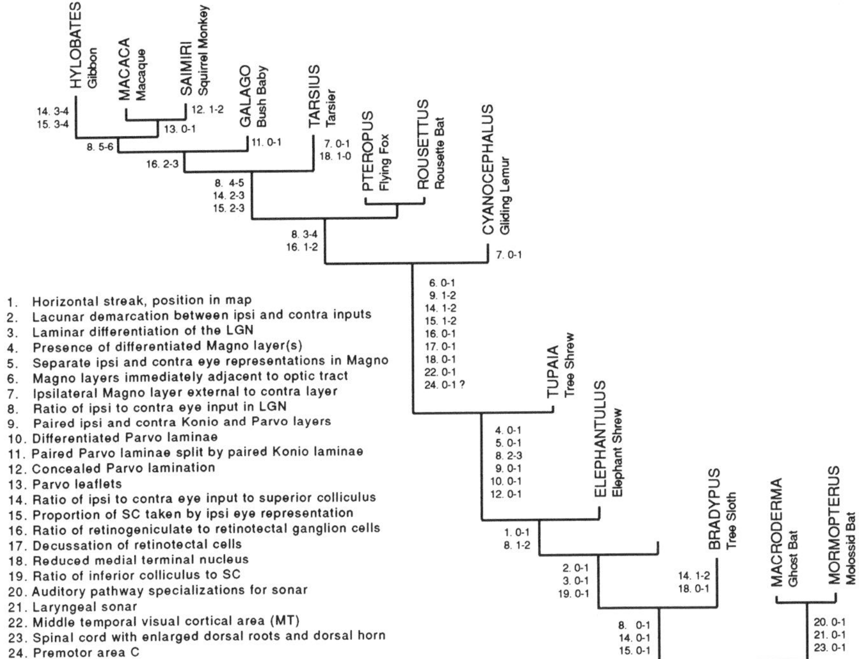

Figure 2. Phylogenetic relationships of bats and other mammalian taxa based on 24 neural characters, and the list of characters (after Pettigrew, 1991a).

Case 2: Fish Sperm

A second example of the use of unconventional characters is that of the work of Jamieson (summarized in 1991) on fish evolution and systematics based on spermatozoa. Sperm structure has long been important in classification of some invertebrate groups, but has not been used much by vertebrate systematists. Jamieson emphasizes ultrastructural characters of sperm and claims to have amplified the field of "spermiocladistics" (Fig. 3). There is indeed great diversity in sperm morphology (Fig. 4), but even the introduction to Jamieson's book calls spermatozoa a "character complex". Much of Jamieson's work is in fact the allocation of sperm ultrastructural apomorphies mapped on cladograms or phylogenetic trees generated from other data and by other workers, notably Lauder and Liem (1983) (Fig. 5). However, Jamieson does occasionally create cladograms based on sperm ultrastructural characters exclusively, but he does not analyze the degree of interdependence of the nearly 70 characters he finds for spermatozoa. I question the power of cladograms based on characters whose validity is suspect. Little or nothing is known about individual variation or morphological development of these characters. I also worry about the effect of introducing some 70 sperm ultrastructural characters into combined databases on computer-assisted methodologies now available. Further, character utility is essential not only at the level of computing the cladogram, but also in obtaining the data. The ability of many systematists to obtain live sperm, and then to have at hand a transmission electron microscope with which to examine ultrastructural detail, may be limited. Further, I question the use of sperm characters in the absence of other data. They may corroborate existing patterns, or if patterns of characters are not congruent with cladograms based on other data, then they may indicate problems of morphological interest. I question the power of cladograms based on characters whose independence is suspect (in other words characters which are likely to be functionally or developmentally linked). On the other hand, this is pioneering work that could provide a new understanding of variation in sperm morphology. It is exactly the exploration of new characters, done in a phylogenetic context, that I and others advocate. I am not

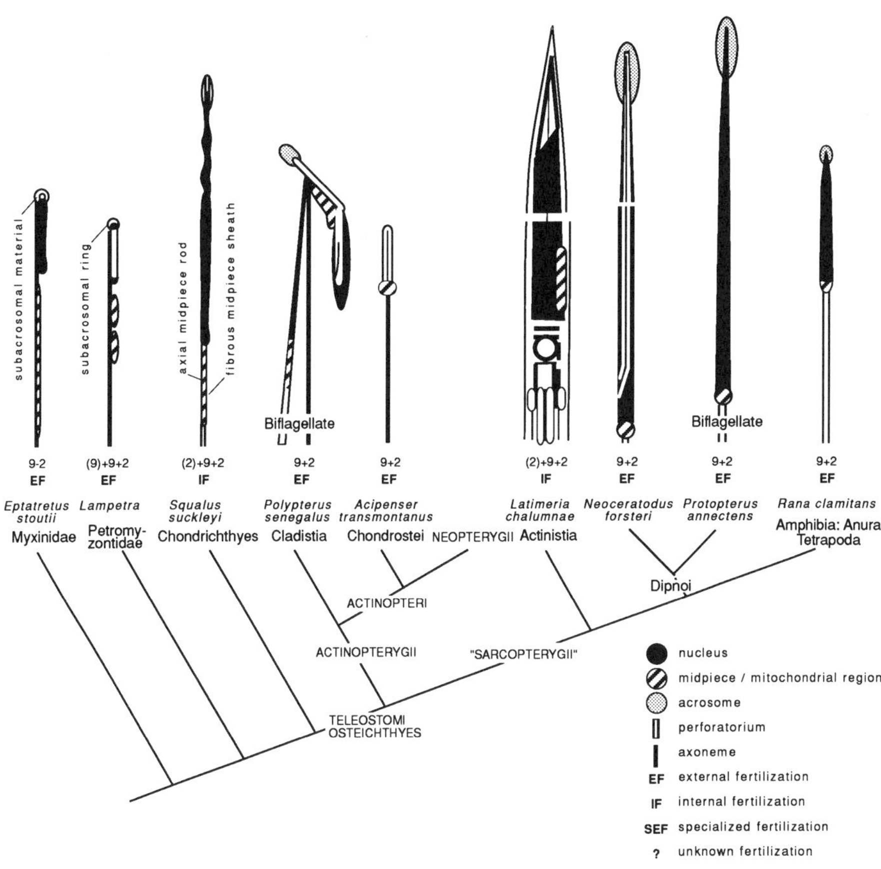

Figure 3. Phylogenetic trends in vertebrate sperm (after Jamieson, 1991).

convinced that we understand these sperm characters adequately, but their further analysis should prove useful. The next step could be to add the sperm data to those that Lauder and Liem (1983) had available, and to re-analyze the phylogenetic relationships.

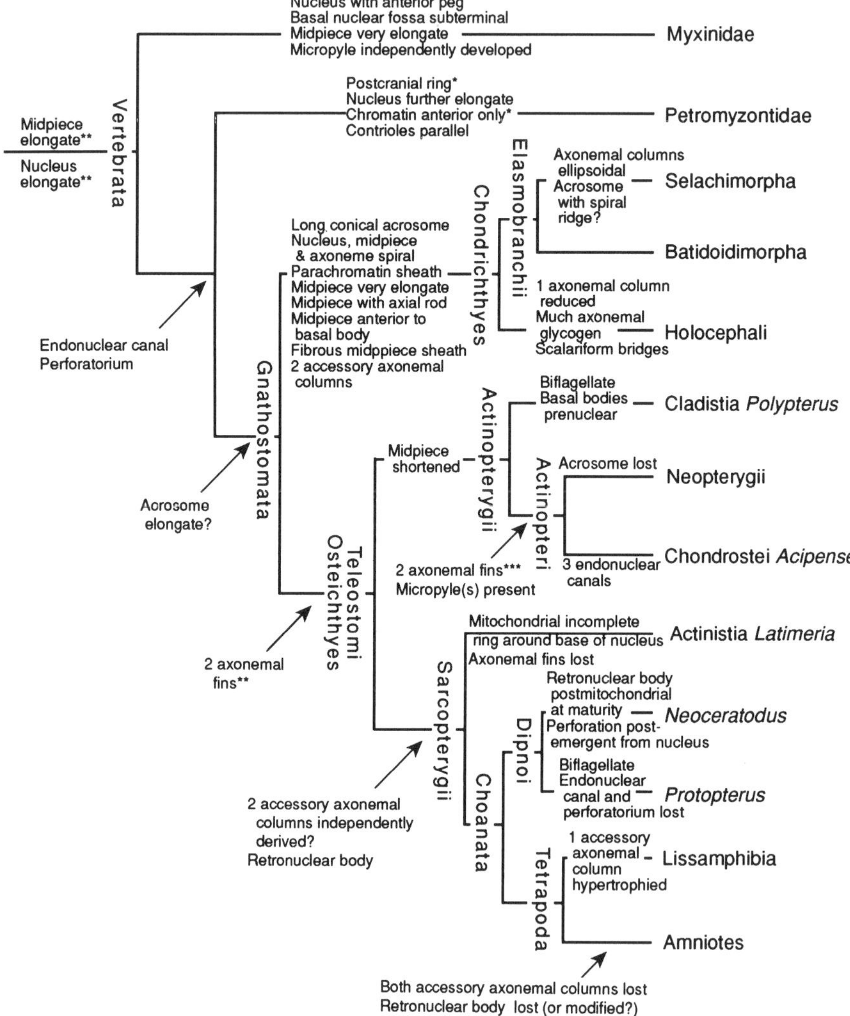

Figure 4. Phylogeny of gnathostomes, indicating variation in sperm morphology (after Jamieson, 1991).

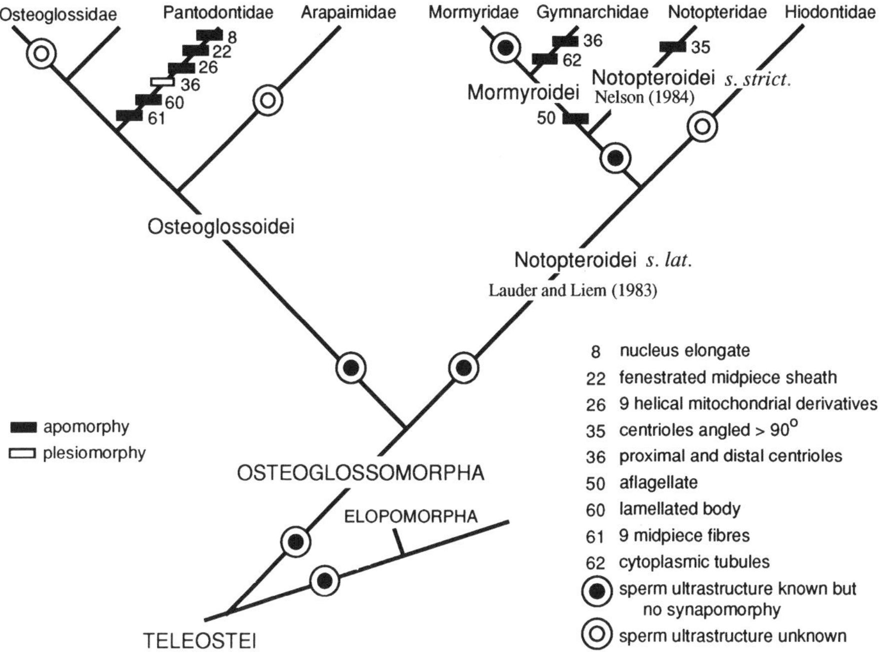

Figure 5. Phylogenetic relationships of Osteoglossomorpha from Lauder and Liem (1983), with apomorphies of spermatozoa mapped on the tree (after Jamieson, 1991).

Case 3: Caecilian Neuroanatomy

My own research attempts the same kind of exploration, but it is combined with the many concerns that I have raised about character analysis and applicability to current phylogenetic methodologies. First, much of my work as an evolutionary morphologist has centered on the Order Gymnophiona of the Class Amphibia, the caecilians. There is a reasonably robust phylogeny for caecilians only at the interfamilial level (Duellman and Trueb, 1986; Nussbaum and Wilkinson, 1989). Relationships within families and genera are less well known. To investigate the evolution of developmental and functional morphology, I must first assess putative patterns of relationship among the taxa I examine. I also try to examine my data to reveal characters of systematic utility, and to understand both the characters and their biological implications in a phylogenetic context. Work in progress illustrates both the positive and negative features that such analyses provide. I have been studying the neuroanatomy of caecilians recently, with a primary goal of better understanding their functional morphology, but also to understand patterns of so-called "regressive" evolution. I have examined the comparative and developmental morphology of eyes (Wake, 1985), ears (Fritzsch and Wake, 1988), olfactory and vomeronasal systems (Billo and Wake, 1987; Schmidt and Wake, 1990), and other special sensory organs, and am now concentrating on the brain and the brainstem motor nuclei, and the peripheral nerves, especially the cranial and anterior spinal nerves (Wake, 1992a, 1993a, 1993b). Two examples will suffice–the morphology of the eye and the morphology of the hypoglossal nerve.

The eyes of all caecilians are overlain by skin, and in some taxa by bone as well. All of the components of accommodation are absent, but there is variability in the way the extrinsic eye muscles, lens, retina, and optic nerve have been lost or modified. However, a retina, an optic nerve, and the optic tectum are all that is necessary for light-dark perception. In addition, all caecilians have modified several eye components to form a new structure, the tentacle, whose lumen enters the vomeronasal organ. Billo and Wake (1987) described the development of the tentacular structure, and the homologies of its parts. The retractor tentaculi is homologous (based on topological and developmental similarity) to the retractor bulbi of frogs and salamanders. The compressor tentaculi is homologous with the levator bulbi, the Harderian gland lubricates the tentacle channel, the tentacle fold is the homolog of the medial part of the lower eyelid, and the tentacle sheath is equivalent to part of the conjunctiva (see Table 1). Thus the eye could have been modified to an organ for olfaction, to stretch the point. In any case, secondary reduction of structures may

Table 1. Homologies of Tentacle Elements (After Billo and Wake, 1987)

Caecilians	Other terrestrial amphibians
Tentacle fold	Lower eyelid
Tentacle aperture	Interpalpebral space
Tentacle sheath	Conjunctival sac
Tentacle sac	Conjunctiva of the eye
Tentacle ducts	Lacrimal ducts
Harderian gland	Harderian gland
M. compressor tentaculi	M. levator bulbi
M. retractor tentaculi	M. retractor bulbi

provide opportunity for morphological innovation. Twelve characters, two with seven states each, of potential taxonomic utility emerged from the comparative data (Table 2).

A second example of this sort of comparative morphology is the examination of the composition of the hypoglossal nerve in caecilians (the nerve that innervates the tongue musculature and some of the hypobranchial muscles). I observed much more variation in the composition of the hypoglossal trunk than I would have expected (Wake, 1992a, 1993a, 1993b) based on the literature and the condition in the outgroups, salamanders and frogs. In frogs, the hypoglossal is composed of fibers of spinal nerve 2; in salamanders it is usually spinals 1 and 2, but in plethodontids it is solely spinal 1. In caecilians, however, the hypoglossal always includes fibers of spinals 1 and 2, and in some taxa spinal 3, and/or the vagus; in several taxa, an occipital nerve also contributes to the hypoglossal. Analysis of the components of the hypoglossal trunk, and its fusions and branching patterns, yielded five

Table 2. Eye Characters (After Wake, 1985, 1993a)

Character 1: Extrinsic musculature: all six present (0); six present, superior oblique attenuate (1a); rectus externus, rectus internus, rectus inferior, inferior oblique present (1b); rectus superior, rectus inferior, rectus internus present (1c); rectus superior, rectus inferior present (1d); rectus inferior present (1e); all absent (1f)

Character 2: Eye under skin (0); eye under skin and bone (1)

Character 3: Optic nerve well developed (0); optic nerve attenuate (1a); optic nerve absent (1b)

Character 4: Vitreous body present (0); vitreous body absent (1)

Character 5: Retina in multiple, well-organized layers (0); retina amorphous (1)

Character 6: Retinal cell number large (greater than 5000) (0); retinal cell number reduced (less than 5000) (1)

Character 7: Number of cell layers in outer cell layer 2 (0); 1-2 (1a); 2-3 (1b)

Character 8: Number of cell layers in inner cell layer 3-4 (0); 3 (1a); 2-3 (1b); 2 (1c); 1 (1d)

Character 9: Number of cell layers in ganglionic cell layer 2 (0); 1-2 (1a); 1 (1b); more than 2 (1c); 0 (1d)

Character 10: Lens round (0); lens spheroid (1a); lens amorphous (1b); lens absent (1c)

Character 11: Lens equal to or greater than 1/2 orbit dia (0); lens less than 1/2 orbit (1a); lens absent (1b)

Character 12: Lens crystalline (0); cellular peripherally, crystalline medially (1a); crystalline peripherally, cellular medially (1b); cellular (1c); cellular, amorphous (1d); rudimentary (1e); absent (1f)

---

potential characters (Table 3), one with eight states.

I illustrate the problems of character interdependence and of the use of limited data sets. I wanted to see what the results of cladistic analysis both of these data alone, and with respect to the phylogenetic hypothesis based on other data, would indicate. I attempted to use characters with little interdependence, but such interdependence is difficult to identify *a priori* (i.e., without reference to congruence). How interdependent or independent are the features of the eye? Based on a combination of embryology and congruence, I suspect that in a fully developed stage, the muscles, lens, retina, covering, etc., are independent. One bit of support for this assumption is that most of these characters do not co-vary; that is, for example, lens modifications (Character 10) do not *appear* to be nonindependent with loss of particular extrinsic muscles (Character 1). The cladogram based on eye data (Fig. 6A), on hypoglossal nerve data (Fig. 6B), and based on a larger neuroanatomical data set that included eye, ear, vomeronasal-olfactory and hypoglossal characters (Fig. 6C) yielded several similarities in pattern, but included some differences. The similarities support the analysis; species within genera cluster together. *Epicrionops* is the sister taxon to all other caecilians based solely on eye data; this is congruent with the hypothesis of family relationships based on other characters. The genera of typhlonectids examined (*Chthonerpeton*, *Nectocaecilia*, *Typhlonectes*) are of an aquatic to semi-aquatic family that inhabits much of South America and is thought to be relatively recently derived and closely related (Wake, 1986). The analysis here does not support a theory of close relationship. Analyses were performed using PAUP (Swofford, 1990). A 50% majority rule consensus tree for the eye characters has a Rohlf's consistency index of 0.865 (Fig. 6A). The 50% majority rule consensus tree for the hypoglossal data is less informative (Fig. 6B), which is not surprising given the limited number of characters and their questionable independence (the Rohlf's CI is 0.304). The typhlonectid genera again do not seem to be closely related and have unexpected sister taxa. Genera in the paraphyletic family Caeciliaidae have relationships that group South American and African taxa in several instances. The

Table 3. Hypoglossal Nerve Characters (After Wake, 1992a, 1993a)

Character 1: Hypoglossal composition: spinals 1 and 2 (0); spinals 1, 2, and 3 (1a); spinals 1 and 2 and occipital (1b); spinals 1, 2, and 3 and occipital (1c); spinals 1, 2, and 3 and ramus X-1 (1d); spinals 1 and 2 and ramus X-1 (1e); spinal 2 predominately and ramus X-1 (1f); spinals 2 and 3 (1g)
Character 2: Occipital absent (0); occipital present (1) Character 3: Distance from atlas to hypoglossal fusion short (adjusted to size of animal) (0); long (1)
Character 4: Length of hypoglossal fusion short (adjusted to size of animal) (0); long (1)
Character 5: Distribution of hypoglossal in tongue bifurcated (0); straight with branches along its length (1a); branched only at tip (1b)

composite cladogram (Fig. 6C) of four sets of neuroanatomical data has a Rohlf's consistency index of 0.818. In this case, more characters give better resolution of the relationships. This cladogram must be viewed with caution, however, for many reasons. For example, of 26 OTU's, only 13 are represented in all four data sets; six species are included in only one of the neuroanatomical data sets. I will now generate new cladograms that include only those taxa for which I have all four kinds of data, and more data will be gathered. In the interim, it is useful to see what the general topography of the tree generated by the composite data set. Again, and even more strongly, species within genera cluster together, except for *Scolecomorphus kirkii*, which consistently does not

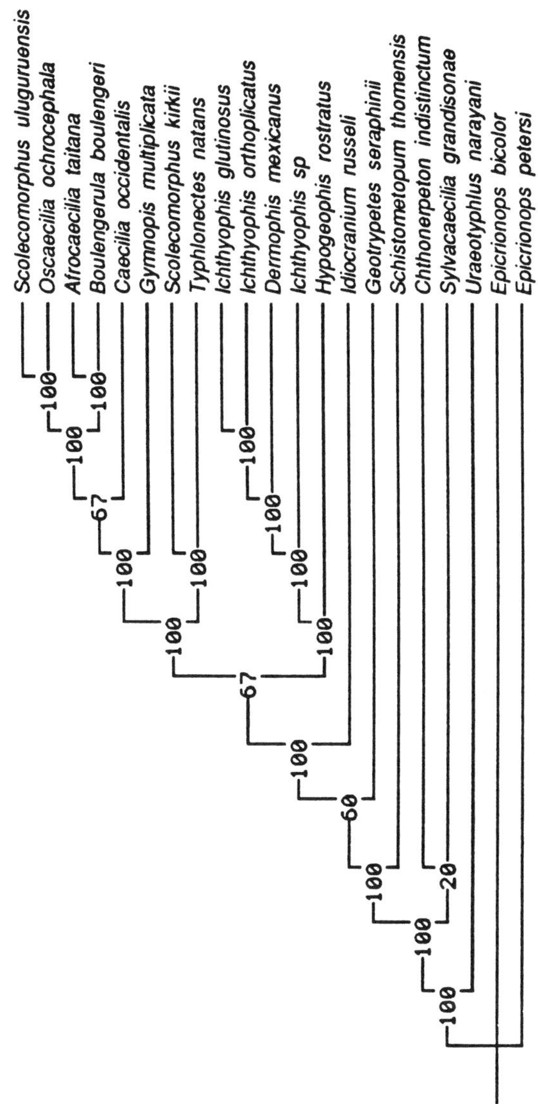

Figure 6A. Cladogram of phylogenetic relationships of gymnophione taxa based on neuroanatomical characters (after Wake, 1993a). 50% majority rule consensus tree based on eye characters. See text for discussion.

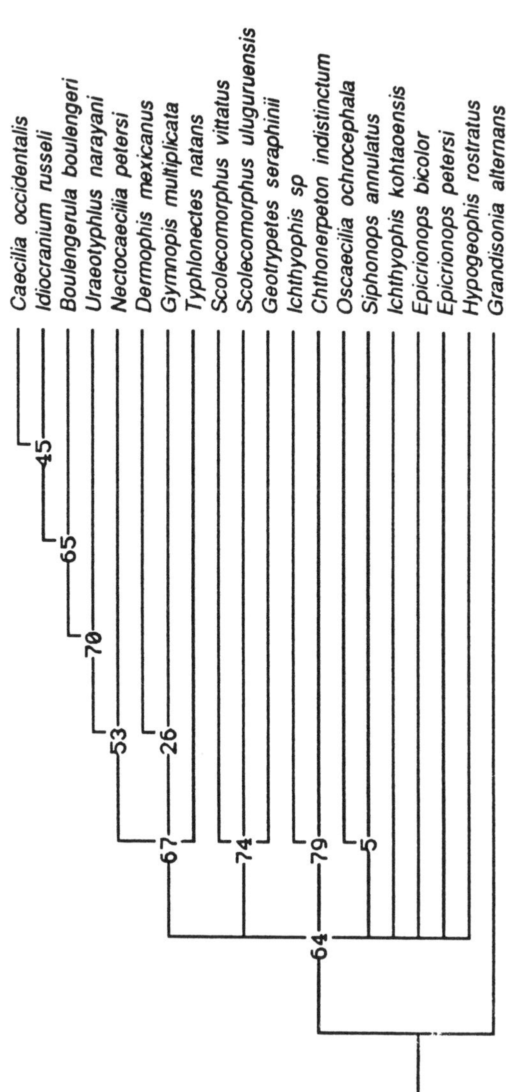

Figure 6B. Cladogram of phylogenetic relationships of gymnophione taxa based on neuroanatomical character (after Wake, 1993a); 50% majority rule consensus tree based on hypoglossal nerve characters. See text for discussion.

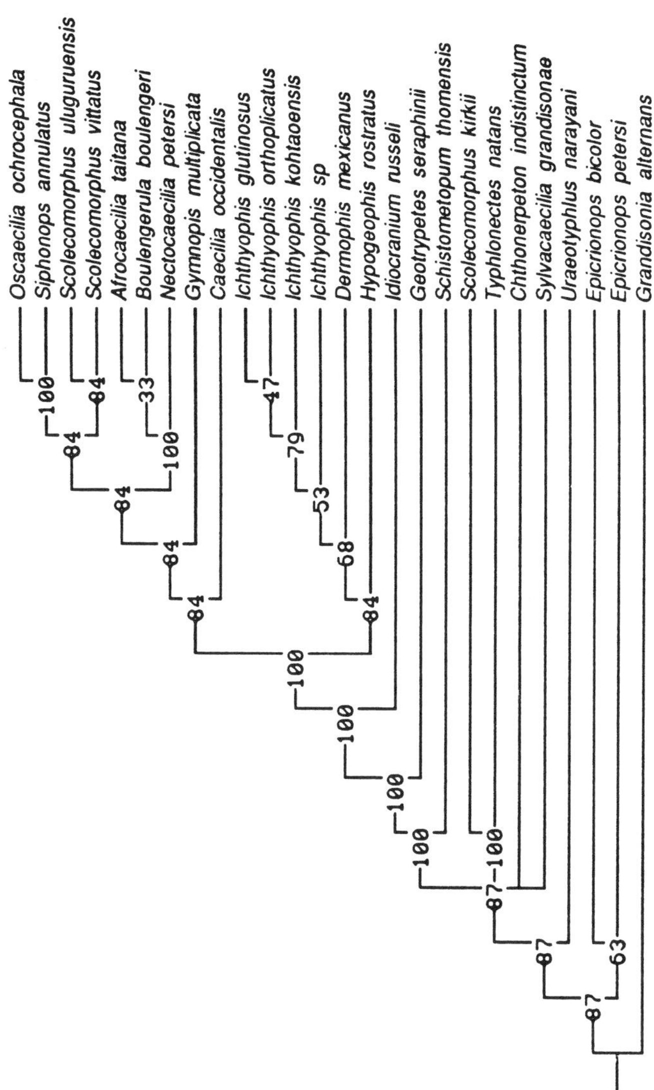

Figure 6C. Cladogram of phylogenetic relationships of gymnophione taxa based on neuroanatomical character (after Wake, 1993a); 50% majority rule consensus tree based on eye, ear, olfactory-vomeronasal, and hypoglossal characters combined. See text for discussion.

group with the other two species of *Scolecomorphus*. *Epicrionops* is the sister taxon to all other caecilians, except *Grandisonia*. The position of *Grandisonia* may be due to the methodology as well as the characters; it is represented in only one small data set (hypoglossal nerve), and all of its characters are plesiomorphic. The east African caeciliaids *Afrocaecilia* and *Boulengerula* are sister taxa, which was not the case for the separate analyses. The relationship of the west African caeciliaids with the Seychellian *Hypogeophis* is consistent among most of the trees. Two genera of typhlonectids are sister taxa, but the third is the sister taxon of the east African caeciliaids; as noted above, the members of the family Typhlonectidae are considered closely related based on other data (Wake, 1986).

When I map the hypoglossal characters on the existing phylogenetic hypothesis of family relationships, I find unexpected patterns of transformation. For example, the occipital nerve contributes to the hypoglossal nerve in derived families of caecilians. Based on the ontogeny and phylogeny of the vertebrate head, this is counter-intuitive, because fishes have larger but variable numbers of spino-occipital nerves (and post-otic head segments), that are reduced, absent, or converted to the true spinals in tetrapods. Therefore the presence of an occipital (it is not a vagus root or a spinal accessory [Wake, 1993b]) that contributes only to the structure of the hypoglossal trunk is an unexpected homoplasy. I therefore continue work to better understand the developmental, comparative, and functional morphology of neural features to further explore the use of neuroanatomical characters as they relate to other morphological characters in phylogenetic analysis. For example, are neuroanatomical features influenced by developmental constraints that differ from those of other systems because they *are* neuroanatomical? Neuroanatomy is subject to ontogenetic repatterning, and it may be that steps in neural ontogeny are better characters for assessing phylogenetic relationships (and pattern and process) than are the features of adult morphology (see Roth et al., 1993, for discussion). I am also concerned about how such characters can be integrated effectively with other morphological data, and with molecular and other data as well. The value of these kinds of studies is that they stimulate analysis of more characters, and in more depth.

Case 4: Caecilian Sperm

Sperm structure poses even more problems than neuroanatomy. I use a silver-staining technique to examine the morphology of sperm taken from preserved animals. I suspect there are some problems with the silver-staining technique as applied to long-preserved material; for example, I would expect a velum on the sperm flagellum in more taxa of caecilians. I consider it important to explore the application of the technique in the search to better understand the biology of the animals, as well as to consider potential characters of systematic utility. Long, recurved sperm heads and acrosomes *appear* to characterize the two genera (and families) (*Epicrionops* [Rhinatrematidae] and *Ichthyophis* [Ichthyophiidae]) considered most primitive of those examined, based on other data, but also *Typhlonectes* and *Chthonerpeton* of the highly derived family Typhlonectidae. (Other taxa have shorter, stouter heads and bodies with blunter acrosomes, but there is variation in these features.) Since I am not convinced of the independence of the potential characters, I have not yet attempted to map them on the family relationships cladogram. I have, however, attempted to begin to understand sperm morphology in another way. I have done a principal components analysis of the meristic data to try to understand what the characters of sperm are (Wake, 1993b). The technique allows me to examine the clustering of taxa based on those data (Fig. 7), as well as size and shape effects. The PCA has produced tantalizing information, as species within genera do cluster, and some genera within families do also, but there are exceptions that I must explore.

These four cases present some of the difficulties inherent in character analysis of "unconventional" morphological data. They illustrate the necessity to understand the biology of the systems and the methods of data retrieval in order to assess character states and polarities. Some understanding of the biology also can serve as a test, of sorts, of character interdependence or independence. The cases also illustrate the contribution that new kinds of morphological data can

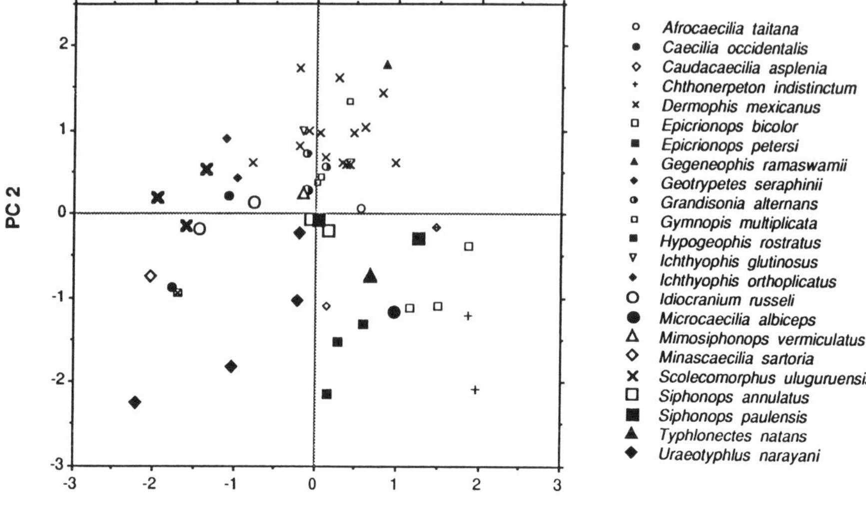

Figure 7. Principal components analysis of meristic data for sperm of 23 gymnophione species representing all families (after Wake, 1993a). See text for discussion.

make to analysis of the nature of homology and of phylogenesis, *if* both data and methodology are well understood.

KINDS OF DATA AND THE GOALS OF PHYLOGENETIC ANALYSIS

Given the many problems with the interpretation and analysis of morphological data for assessment of both systematic pattern and evolutionary process, the use of molecular data to assess relationships and to generate cladograms is of great interest to many biologists. Perhaps the most extreme view of the analysis of morphological and molecular data (DNA data in particular) to assess phylogenetic relationships and patterns of evolution is that held by the structuralists' school. In contrast to many systematists, who believe that molecular data are best used either in conjunction with morphological data or as a form of reciprocal data for analysis of relationships, structuralists state that a phylogeny should be based on molecular data alone. The structuralists assert that the basic assumption of phylogenetic systematics, that classification on the basis of morphology should be congruent with genealogy, is invalid. Ho (1984, 1988, 1989, 1990) states that genealogy is best traced by comparing as many DNA sequences as possible, preferably entire genomes. Ho considers DNA sequences a much more reliable guide to genealogy, since the data are genically based, whereas morphology is also influenced by a diversity of factors. Ho therefore concludes that there is no necessity for there to be complete congruence between morphological data and genealogy. She maintains that genealogy is only one aspect of the relationship between organisms, and investigation of the taxonomy of forms is distinct from the reconstruction of genealogy. A "rational morphology", and a "rational taxonomy" of form, to use Ho's terms, are possible because morphological changes follow natural rules of order; but organisms do not evolve rationally because of the creative, contingent nature of evolution (Webster and Goodwin, 1982, 1984; Ho, 1988, 1989, 1990). These structuralists, in addition to denying the Darwinian paradigm as useful, are really asking that sets of form (e.g., morphologies) be analyzed as transformations of dynamical structure according to developmental programs, not dissimilar to the strategy advocated by Wagner (1989) and others. The structuralists see that this approach will establish biological homology, and that genealogy, as indicated above, is a separate analysis.

I find the structural approach useful because many of the questions I raised in this paper can be better answered if the parts of these questions are separated. However, a comprehensive approach to the study of phylogenetic patterns and evolutionary process must take into account all the data we can obtain, and requires rigorous methods of analysis. New methodologies, and modifications of current (and even classical or traditional) methods, as well as new kinds of data, are essential to developing better understanding of patterns and process of phylogenesis and of evolution. I (Wake, 1990, 1992b, and herein) simply join many, perhaps the majority, of evolutionary morphologists and systematists who conclude that, as expressed by Fisher (1985), "A comprehensive understanding of form and form-change requires that this be integrated with the perspective offered by studies of development, genetics, phylogenetic history, and external perturbations acting on the system."

Acknowledgments

I thank Olivier Rieppel and others associated with the Field Museum Spring Systematics Symposium for the invitation to participate in this discussion of Systematic Process. I appreciate the constructive comments of Hugh Griffith, Olivier Rieppel, David Wake, and anonymous reviewers, on the manuscript. I thank the National Science Foundation for support of my research in evolutionary morphology and the analysis of morphological data.

References

Baker, R. J., M. J. Novacek and N. B. Simmons. 1991. On the monophyly of bats. *Systematic Zoology*, 40: 216-231.

Billo, R., and M. H. Wake. 1987. Tentacle development in *Dermophis mexicanus* (Amphibia: Gymnophiona: Caeciliidae). *Journal of Morphology*, 192: 101-111.

Brooks, D. R., and D. A. McLennan. 1991. *Phylogeny, Ecology, and Behavior*. Chicago: University of Chicago Press.

Duellman, W. E., and L. Trueb. 1986. *Biology of Amphibians*. New York: McGraw-Hill Book Co.

Felsenstein, J. 1988. The detection of phylogeny. In D. L. Hawksworth (Ed.), *Prospects in Systematics*, 112-127. Oxford: Clarendon Press.

Fisher, D. C. 1985. Evolutionary morphology: beyond the analogous, the anecdotal, and the ad hoc. *Paleobiology*, 11: 120-138.

Fritzsch, B., and M. H. Wake. 1988. The inner ear of gymnophione amphibians and its nerve supply: a comparative study of regressive events in a complex sensory system (Amphibia: Gymnophiona). *Zoomorphology*, 108: 201-217.

Gans, C., and R. G. Northcutt. 1983. Neural crest and the origin of vertebrates: a new head. *Science*, 220: 268-274.

Goodwin, B. C. 1984. A relational or field theory of reproduction and its evolutionary implications. In M.-W. Ho and P. T. Saunders (Eds.), *Beyond Neo-Darwinism: An Introduction to the New Evolutionary Paradigm*, 219-241. London: Academic Press. 376 pp.

Hickman, C. S. 1988. Analysis of form and function in fossils. *American Zoologist*, 28: 775-793.

Ho, M.-W. 1984. Where does biological form come from? *Revista di Biologia*, 77: 147-179.

Ho, M.-W. 1988. How rational can rational morphology be? A post-Darwinian rational taxonomy based on a structuralism of process. *Revista di Biologia*, 81: 11-55.

Ho, M.-W. 1989. A structuralism of process: towards a post-Darwinian rational morphology. In B. A. Goodwin, A. Sibatani and G. Webster (Eds.), *Dynamic Structures in Biology*, 31-48. Edinburgh: Edinburgh University Press.

Ho, M.-W. 1990. An exercise in rational taxonomy. *Journal of Theoretical Biology*, 147: 43-57.

Jamieson, B. G. M. 1991. *Fish Evolution and Systematics: Evidence from Spermatozoa*. Cambridge: Cambridge University Press.

Kluge, A. G. 1989. A concern for evidence and a phylogenetic hypothesis of relationships among *Epicrates* (Boidae, Serpentes). *Systematic Zoology*, 38: 7-35.

Lauder G. V. 1990. Functional morphology and systematics: studying functional patterns in an historical context. *Annual Reviews of Ecology and Systematics*, 21: 317-340.

Lauder, G. V., and K. F. Liem. 1983. The evolution and

interrelationships of the actinopterygian fishes. *Bulletin of the Museum of Comparative Zoology*, 150: 95-197.

Lauder, G. V., and K. F. Liem. 1989. The role of historical factors in the evolution of complex organismal functions. In D. B. Wake and G. Roth (Eds.), *Complex Organismal Functions: Integration and Evolution in Vertebrates*, 63-78. Chichester: John Wiley.

Liem, K. F. 1989. Functional morphology and phylogenetic testing within the framework of symecomorphosis. *Acta Biotheoretica*, 27: 119-131.

Nelson, G. F. 1970. Outline of a theory of comparative biology. *Systematic Zoology*, 19: 373- 384.

Nelson, G. F. 1989. Cladistics and evolutionary models. *Cladistics*, 5: 257-289.

Northcutt, R. G., and C. Gans. 1983. The genesis of neural crest and epidermal placodes: a reinterpretation of vertebrate origins. *Quarterly Review of Biology*, 58: 1-28.

Nussbaum, R. A., and M. Wilkinson. 1989. On the classification and phylogeny of caecilians (Amphibia: Gymnophiona), a critical review. *Herpetological Monographs*, 3: 1-42.

Pettigrew, J. D. 1991a. Wings or brain? Convergent evolution in the origins of bats. *Systematic Zoology*, 40: 199-216.

Pettigrew, J. D. 1991b. A fruitful, wrong hypothesis? Response to Baker, Novacek, and Simmons. *Systematic Zoology*, 40: 231-239.

Pettigrew, J. D., B. G. M. Jamieson, S. K. Robson, L. S. Hall, K. I. Macnally and H. M. Cooper. 1989. Phylogenetic relationships between microbats, megabats and primates (Mammalia: Chiroptera and Primates). *Philosophical Transactions of the Royal Society of London*, Series B, *Biological Science*, 325: 489-559.

Rieppel, O. 1988. *Fundamentals of Comparative Biology*. Basel: Birkhauser Verlag.

Roth, G., K. C. Nishikawa, C. Naujoks-Manteuffel, A. Schmidt and D. B. Wake. 1993. Paedomorphosis and simplification in the nervous system of salamanders. *Brain, Behavior, and Evolution*, 42: 137-170.

Roth, V. L. 1984. On homology. *Biological Journal of the Linnean Society*, 22: 13-29.

Roth, V. L. 1988. The biological basis of homology. In C. J. Humphries (Ed.), *Ontogeny and Systematics*, 1-26. New York: Columbia University Press.

Schmidt, A., and M. H. Wake. 1990. The olfactory and vomeronasal system of caecilians (Amphibia: Gymnophiona). *Journal of Morphology*, 205: 255-268.

Simmons, N. B., M. J. Novacek and R. J. Baker. 1991. Approaches, methods, and the future of the chiropteran monophyly controversy: a reply to J. D. Pettigrew. *Systematic Zoology*, 40: 239-243.

Swofford, D. L. 1990. Phylogenetic analysis using parsimony (PAUP), version 3.0. Urbana: Illinois Natural History Survey.

Van Valen, L. 1982. Homology and causes. *Journal of Morphology*, 173: 305-312.

Wagner, G. P. 1989. The origin of morphological characters and the biological basis of homology. *Evolution*, 43: 1157-1171.

Wake, D. B. 1991. Homoplasy: the result of natural selection, or evidence of design limitations? *American Naturalist*, 138: 543-567.

Wake, D. B., and A. Larson. 1987. Multidimensional analysis of an evolving lineage. *Science*, 238: 42-48.

Wake, M. H. 1985. The comparative morphology and evolution of the eyes of caecilians (Amphibia: Gymnophiona). *Zoomorphology*, 105: 277-295.

Wake, M. H. 1986. A perspective on the systematics and morphology of the Gymnophiona (Amphibia). *Memoires de la Societe Zoologique de France*, 43: 21-38.

Wake, M. H. 1990. The evolution of integration of biological systems: an evolutionary perspective through study of cells, tissues, and organs. *American Zoologist*, 30: 897- 906.

Wake, M. H. 1992a. Patterns of peripheral innervation of the tongue and hyobranchial apparatus in caecilians (Amphibia: Gymnophiona). *Journal of Morphology*, 212: 37-53.

Wake, M. H. 1992b. Morphology, the study of form and function, in modern evolutionary biology. In D. Futuyma and J. Antonovics (Eds.), *Oxford Surveys in Evolutionary Biology*, 9: 289-346. New York: Oxford University Press.
Wake, M. H. 1993a. Non-traditional characters in the assessment of caecilian phylogenetic relationships. *Herpetological Monographs*, 7: 42-55.
Wake, M. H. 1993b. Evolutionary diversification of cranial and spinal nerves and their targets in the gymnophione amphibians. *Brain, Behavior and Evolution. Acta Anatomica*, 148: 160-168.
Webster, G., and B. C. Goodwin. 1982. The origin of species: a structuralist approach. *Journal of the Society of Biological Structure*, 5: 15-42.
Webster, G., and B. C. Goodwin. 1984. A structuralist approach to morphology. *Revista di Biologia*, 77: 503-531.
Wiley, E. O. 1981. *Phylogenetics: The Theory and Practice of Phylogenetic Systematics*. New York: John Wiley and Sons.

8

The Phylogeny of Development and the Origin of Homology

Neil H. Shubin

Department of Biology
University of Pennsylvania
Philadelphia, Pennsylvania 19104

Abstract. Analyses of ontogeny have traditionally played important roles in the study of morphological homology. The conflation of the criteria used to formulate, test, and mechanistically explain hypotheses of homology can lead to difficulties in understanding the relationships between ontogeny and homology. Historically, similarity of ontogenetic patterns and mechanisms has been used as a major test of homology because detailed convergence of ontogenetic pattern and mechanism has been considered unlikely. Ontogenetic similarities from a single system need not, however, overturn hypotheses of homology based on other anatomical systems and characters. Comparison of ontogenies provides important information necessary to formulate hypotheses of homology in that it informs about the biological basis of the conservatism and individuation of characters. Phylogenetic hypotheses can provide important information on the evolution of ontogenies—information that would be missed if developmental transformations alone were used as the test of homology. Studies of development assist in the formulation of hypotheses of homology and their mechanistic explanation. Phylogenetic analysis provides tests of these hypotheses and enables the historical analysis of ontogenetic transformations.

Copyright © 1994 by Academic Press, Inc.
All rights of reproduction in any form reserved.

THE THEME OF THIS VOLUME REFLECTS a major issue in the study of homology. The analysis of homology informs, and is informed by, the study of biological processes. The recognition of homologous features is a key step in the analysis of phylogenetic patterns and, in addition, our ability to formulate robust hypotheses of homology reflects meaningful properties of evolutionary processes. Studies of ontogeny can inform about homology because homologous features emerge by the genetic and environmental modification of inherited developmental systems (Müller, 1990; Müller and Wagner, 1991). The role of these biological mechanisms in the analysis of homology has been extensively discussed over the past 150 years (see reviews in: Bock, 1977; Patterson, 1982, 1988; Goodwin and Trainor, 1983; Roth, 1984, 1988, 1991; Rieppel, 1980, 1988; Wagner, 1989a, 1989b; de Pinna, 1991; Haszprunar, 1992). At issue in these discussions are the criteria used in the formulation, validation, and mechanistic explanation of hypotheses of homology. In this chapter, I review the relationship between developmental data and the analysis of homology to explore the use of pattern and process in evolutionary morphology.

HOMOLOGY CONCEPTS

The notion of homology describes continuity and conservatism in the evolution of diversity (Van Valen, 1982; Roth, 1991). Many different notions of homology have been proposed over the past 150 years. These approaches differ in biological units used for comparison and in the phenomena that they attempt to explain. Recent discussions have focused on two closely related concepts: "biological homology" (Roth, 1984, 1988, 1991; Wagner, 1989a, 1989b; Minelli and Peruffo, 1991; Haszprunar, 1992) and "phylogenetic homology". These two concepts are not at odds, but are complementary as discussed below.

Phylogenetic Homology

There are many definitions of phylogenetic homology (see references in: Bock, 1977; Rieppel, 1980, 1988; Patterson, 1982, 1988; de

Pinna, 1991). The concept of phylogenetic homology is a historical approach to the continuity of information and, as a consequence, involves the hierarchical comparison of features in different taxa. Two features are considered phylogenetically homologous if they are derived from the same feature in a common ancestor. In the framework of cladistics, homology has been equated with synapomorphy (Eldredge and Cracraft, 1980; Patterson, 1982). This approach to homology has also been termed a taxic one, because these phylogenetic homologies imply a hierarchy of groups (Eldredge and Cracraft, 1980; Patterson, 1982, 1988; Rieppel, 1988; de Pinna, 1991).

To discuss phylogenetic homology we must distinguish between the criteria used to formulate, test, and mechanistically explain hypotheses of homology. The conflation of these three goals has led to numerous difficulties. There is no constraint on the means used to formulate hypotheses of homology (de Pinna, 1991). The analysis of biological processes (e.g., ontogeny) provides legitimate criteria for the formulation of hypotheses of homology. Many early studies attempt to outline the rules for formulating hypotheses of homology by identifying "key" features that provide special information about morphological homology and phylogenetic relationships. Similarly, transformational theories (such as Dollo's Law, Haeckel's Biogenetic Law, and Morse's Law, see below) were also used in the identification of homologous features. *A priori* notions of transformation or homology can bias phylogenetic hypotheses by influencing the recognition and polarity of characters. Whereas these theories of transformation are legitimate criteria for the formulation of hypotheses of homology, they neither test nor validate hypotheses themselves. The ability to test theories of transformation must rely on noncircular criteria. Generalizations about phylogenetic transformations cannot be used to both recognize and test hypotheses of homology and phylogeny. Transformational notions of evolutionary change are validated by taxic approaches to homology, not vice versa.

The methodological constraint in the analysis of homology lies in the test or validation of proposals of homology. The recognition and test of phylogenetic homology is a two-step procedure: hypotheses of homology are proposed and are later tested by phylogenetic analysis (de Pinna, 1991) in the hope of finding congruence and a hierarchical (and parsimonious) summary of data (pattern). The proposal of a hypothesis of homology carries implicit phylogenetic information because it is as much a statement about character evolution as it is a hypothesis of the phylogenetic relationships of the taxa involved. That is, when features are deemed homologous, an explicit prediction is made about the phylogenetic relationships of the taxa under comparison. If a particular hypothesis of homology is not supported by phylogenetic analysis, then the original hypothesis of homology may be in error. The phylogenetic tests of a particular hypothesis of homology rely on congruence with other proposed homologies or synapomorphies (Patterson, 1982, 1988; de Pinna, 1991).

Phylogenetic homology is an explicit statement about common descent and is only indirectly related to similarity. Not all similarities are homologous and, conversely, homologous features may be very dissimilar. Similarity has been cited as being both a recognition criterion for homology and as a test of hypotheses of homology. The use of similarity as a clue in the recognition of homologous features is not controversial. Similarity alone is an exceedingly weak test of an hypothesis of homology (Bock, 1977). The test of phylogenetic homology is congruence with other proposed homologies, not phenotypic similarity.

Iterative Homology

Iterative or serial homology is homology between parts of the same organism. A commonly cited example is foliage on an individual plant (Wagner, 1989b).

The implied consequence of homology entails biological explanation for morphological continuity and change. A host of biological process theories including (in part) stabilizing selection, genetic and epigenetic constraints, and lack of additive genetic variability can explain evolutionary conservatism. The mechanistic explication of conservatism involves comparisons at intra- and inter-individual levels, and the "biological homology concept", described below, attempts to provide an approach to unify these different types of comparison under a single conceptual framework.

Biological Homology

The notion of "biological homology" is broader than that of phylogenetic homology or iterative homology. Structures within the same individual may be considered to be "biologically" homologous if they are produced by the same generative mechanisms that produce similar patterns of individuality and developmental constraints. Wagner (1989a, 1989b) intentionally does not distinguish between serial homologies and other types of homology because he believes they share the same biological cause. As a consequence, those characteristics that serve as synapomorphies in phylogenetic hypotheses (i.e., phylogenetic homologies) are a subset of the broader class of Wagner's biological homologies.

Recent attention to the concept of homology has focused on the causal basis of morphological conservatism. Common descent alone is not causal to homology; needed are invariant features of organisms. Van Valen (1982, p. 305) has proposed that homology is "correspondence caused by a continuity of information". In this context, "information" is used to describe the genetic, epigenetic, and other mechanisms that serve to specify phenotypes. The manner by which this "information" is transmitted defines the types of comparisons that are possible. This approach makes the distinction between iterative (serial) homology and taxic homologies unnecessary because taxic homologies are but one expression of this type of homology.

The central issue in the causal analysis of homology is the recognition of the classes of biological processes that cause the invariance, lability, and individuality of morphological structures. Biological homologies are features that share a set of specific developmental constraints. The concept of "Biological Homology" has been defined by Wagner (1989a, p. 62) as follows: "Structures from two organisms or from the same individual are homologous if they share a set of developmental constraints, caused by locally acting self-regulating mechanisms of organ differentiation. These structures are thus developmentally individualized parts of the phenotype".

The recognition of biological homologs, and the tests of these hypotheses are based upon the similarity of genetic and epigenetic interactions, rather than on congruence with other proposed homologies. A hypothesis of biological homology is a statement about the manner in which a set of features is able to vary. If features share a set of conserved developmental constraints they will maintain similar norms of reaction and will have similar patterns of biased variation. The tests of the biological homology concept rely on experimental manipulations of epigenetic processes, the analysis of variability in populations, and phylogenetic data.

Hypotheses of biological homology will contain a phylogenetic "signal" if the developmental mechanisms that are involved in the formation of a biological homolog are so complex that they are unlikely to have evolved twice. As a consequence, these biological homologs should be congruent with robust phylogenetic hypotheses. The problem with this approach is that we have no *a priori* way of knowing how many times a structure, or a set of epigenetic interactions, has evolved. There is no rigorous way in which biological homology concepts can be used to distinguish between phylogenetic homology and homoplasy. This does not imply that the biological approach to homology carries no phylogenetic information. Phenetic analyses of biological homologies can provide important hypotheses of character identification. Phylogenetic tests of these hypotheses can reveal the extent to which these developmental processes are conserved.

Biological and phylogenetic hypotheses of homology may not be congruent. That is, experimental and descriptive studies of development may suggest that a particular structure is homologous, but phylogenetic information may suggest multiple independent origins for these biological homologs. The cases of incongruence between biological homologs and phylogenetic ones can result from:

1) incorrect hypotheses of biological homology
2) incorrect phylogenetic hypotheses
3) homoplasy of form-generating mechanisms

How do we differentiate between these three alternatives? To integrate these approaches we need to discuss how biological and phylogenetic concepts can be used to test one another.

Developmental Approaches to Homology

Developmental data has played a variety of roles in the analysis of morphological homology. Developmental studies of morphological homology have such a long history in evolutionary biology that this subject cannot be reviewed fully here. Below, I focus on two perspectives that continue to have a pervasive influence.

Recapitulation: Homology and Polarity

Classical embryological studies used the principle of recapitulation both to identify homologous characters and to polarize them. Strict notions of Haeckelian recapitulation either implicitly or explicitly assume a conservatism of early developmental stages. This approach was used to: (1) compare corresponding features during the ontogeny of a single individual, and (2) compare both ontogenies and adult features in different taxa. Similarly, all methods (including those that do not rely on Haeckelian interpretations) that utilize developmental data to polarize characters (the ontogenetic method, *sensu* Nelson, 1978; Patterson, 1982) implicitly suppose that ontogeny can also be used to recognize homologs. This approach is only possible if one could homologize a single structure during the ontogeny of an individual and extend this "ontogenetic homology" (*sensu* Haszprunar, 1992) to comparisons between taxa.

A major example of this approach comes from studies of tetrapod limbs (Fig. 1). Phylogenetic comparisons, using both extant and fossil forms, suggested that the limbs of many vertebrate taxa arose by a reduction or loss of different components of the limb skeleton

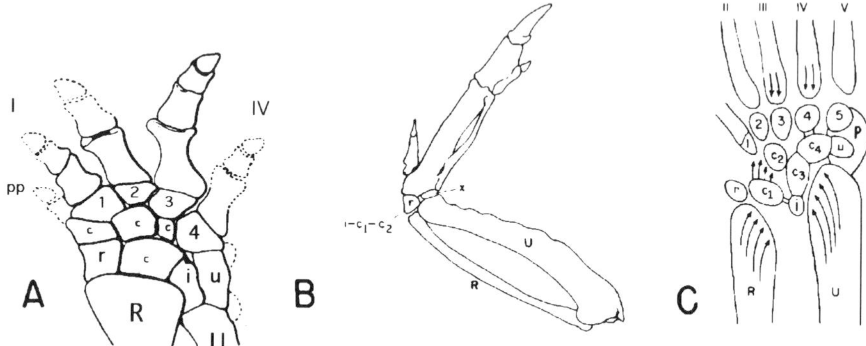

Figure 1. (A) The forelimb of *Eryops* reveals the general condition of the tetrapod limb. These generalized limbs have numerous carpal bones in the manus. These elements have been identified as: I-IV, digits one through four; 1-4, distal carpals 1-4; c, centralia (up to four are present primitively); i, intermedium; pp, pre-pollex; r, radiale; R, radius; u, ulnare; U, ulna. (B) The derived condition of many tetrapod limbs (as in the wing of *Gallus*, figured here) is the result of a reduction in the number of bones in the manus and pes. (C) Montagna (1945) and other recapitulationists have proposed that early development of the chick wing parallels phylogenetic history. Montagna identified anlagen of numerous carpal bones during the development of the chick wing and suggested that these elements are homologous to those seen in primitive tetrapod limbs. These elements are lost in later development by the processes of differential growth and fusion of anlagen (arrows). Drawings not to scale.

(Hinchliffe and Johnson, 1980). The early development of tetrapod limbs was seen to recapitulate the exact number of limb elements of primitive fossil taxa (Fig. 1 and Holmgren, 1955). These comparative studies of development provided a framework for the analysis of the homology of the elements of the limb skeleton. The similarity of patterns of limb development in many taxa suggested hypotheses of homology despite the fact that the adult bones may have shared few phenotypic similarities.

Recent embryological studies have contradicted a strict application of this view. Developing limbs do not recapitulate the exact number of elements of fossil taxa (Hinchliffe, 1977, 1989, 1991; Hinchliffe and Griffiths, 1983; Shubin and Alberch, 1986). Although reduction or loss of elements occurs during evolution and development, there is no archetypal arrangement observed during early developmental stages. Derived features (including synapomorphies of higher level taxa) can be seen very early in development (Shubin, 1991). Tetrapod limbs share many features of limb development but these shared features do not suggest either von Baerian or Haeckelian recapitulation.

Structuralist Perspectives

The study of the dynamics of developmental systems has led to critiques of the notion of homology altogether (Goodwin and Trainor, 1983). Small changes in developmental parameters can produce coordinated changes across the entire organism. Goodwin and Trainor argue that the recognition of homologous features involves the atomism of phenomena that are part of integrated dynamic systems. Individual anatomical features, to Goodwin and Trainor, may not have an independent identity. This lack of individuality is a product of global morphogenetic mechanisms that cannot necessarily be atomized into smaller discrete processes.

Tetrapod limb reduction serves as an example of the structural approach to comparative biology. Comparative morphologists have proposed homologies between different digits of tetrapod limbs. These homologies result from the comparison of phalangeal formulae of the limbs of different tetrapods. The loss of a digit, or a single phalanx, is a phylogenetic pattern seen independently in many tetrapod taxa. When comparative morphologists compare a reduced, four-digited limb with a primitive pentadactyl pattern, they propose that a specific digit (usually either digit one or five) has been lost. This approach implicitly assumes that the different digits have an independent phylogenetic identity and that this individuality can be used to atomize the limb skeleton into discrete parts. Goodwin and Trainor (1983) suggest that a developing limb arises within a developmental field that is controlled by globally acting developmental parameters. Evolutionary change, in this perspective, is the consequence of the perturbation of these global parameters. Consequently, Goodwin and Trainor propose that individual digits do not have a discrete developmental identity. The act of homologizing individual digits involves, then, an artificial atomism that has no bearing in developmental biology.

The biological homology concept builds on the structural perspective proposed by Goodwin and Trainor (1983). Goodwin and Trainor clearly accept that the limbs of tetrapods are homologous as paired appendages of vertebrates. If they accepted purely global mechanisms of morphogenesis, then they could not even compare individual limbs of different tetrapods (Shubin and Alberch, 1986). Wagner (1989a, 1989b) suggests that the most useful comparisons are between developmentally individuated parts of the phenotype. This approach entails circumscribing phenotypically meaningful limits to the developmental integration of structures. The degree to which developmental systems are individuated dictates the resolution at which homologies can be drawn and the extent to which Goodwin and Trainor's notion applies to morphological evolution (Shubin and Alberch, 1986; Wagner, 1989a, 1989b; Minelli and Peruffo, 1991). Neither phylogenetic nor developmental data support a strict application of Goodwin and Trainor's view. The strongest argument against a strict interpretation of Goodwin and Trainor's notion comes from the analysis of transformation series that depict the directional loss or gain of a single limb bone (Fig. 2). In these cases, one bone is modified with no phenotypic modification of other aspects of the limb. If, in these morphoclines, a single bone is lost and the remaining bones are also altered, the reconstruction of homology may not be possible. The fact that we can make highly corroborated hypotheses of homology using discrete features provides important information on the developmental individuation of morphological structures.

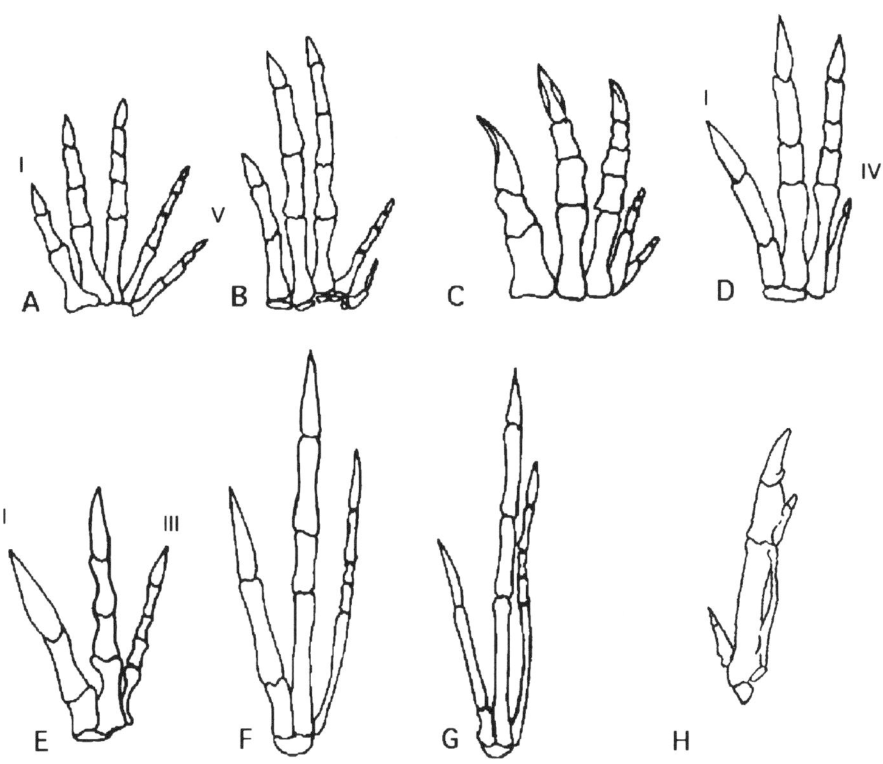

Figure 2. The manus of selected archosaurs: (A) *Crocodylus sp.* (Crocodylomorpha), (B) *Heterodontosaurus* (Ornithischia), (C) *Massospondylus* (Sauropodomorpha), (D) *Syntarsus* (Ceratosauria), (E) *Allosaurus* (Carnosauria), (F) *Deinonychus* (Dromaeosauria), (G) *Archaeopteryx* (Aviale), (H) *Gallus* (Aviale). The preaxial border of each limb is on the left side of the figure.

THE HOMOLOGY OF THE DIGITS OF THE AVIAN WING: CONFLICT BETWEEN BIOLOGICAL AND PHYLOGENETIC APPROACHES TO HOMOLOGY?

The homology of the digits of the avian limb has been cited as an example where biological and phylogenetic approaches to homology are in conflict (see review in Hinchliffe and Hecht, 1984). Several interpretations of the homology of the avian digits have been debated over the past 170 years. The three digits of the wing of modern birds have been identified as being homologous to digits (in these discussions, digits are represented by Roman numerals I-V and phalangeal formulae follow the designations discussed by Padian, 1992): I-II-III, II-III-IV, I-II-IV or III-IV-V of other archosaurs. The latter two interpretations are no longer seriously entertained and will not be discussed here. Hinchliffe and Hecht (1984) argue that the I-II-III and the II-III-IV hypotheses result from the analysis of qualitatively different types of data: workers who give primacy to paleontological data support the I-II-III view whereas those who include embryological information in their analysis tend to support the II-III-IV hypothesis. This disparity, however, is not solely the result of the use of different types of data. These two interpretations are as much the result of different approaches to the analysis of homology as they are due to the use of different types of data.

I-II-III Theories

Some embryologists and many paleontologists have proposed that the digits of the avian wing are homologous to digits I, II, and III of diapsids (Meckel, 1825; Gegenbaur, 1864; Huxley, 1868; Rosenberg, 1873; Jeffries, 1881; Parker, 1888; Zehntner, 1890; Broom, 1906; Siegelbauer, 1911; Abel, 1912; Heilmann, 1927; Steiner, 1922; Romer, 1956; Ostrom, 1976; Gauthier, 1986). Several lines of evidence have been used to support this view: (1) comparison of the phalangeal formula of *Archaeopteryx* with that of other archosaurs, (2) phylogenetic analysis of digital reduction in theropod dinosaurs, and (3) strict application of Haeckel's Biogenetic Law.

The most widely cited support of the I-II-III hypothesis comes from the comparison of the manus of *Archaeopteryx* with the manus of archosaurs (e.g., Gauthier, 1986). The phalangeal formula of the manus of *Archaeopteryx* is 2-3-4-x-x (Fig. 2). The phalangeal formula of primitive archosaurs is a matter of debate, but Romer (1956) hypothesizes it to be 2-3-4-5-3. The three digits of *Archaeopteryx* have similar phalangeal formulae as digits I, II, and III of primitive pentadactyl archosaurs (Fig. 2). If the digits of *Archaeopteryx* are I-II-III, and have a phalangeal formula of 2-3-4-x-x, then the most parsimonious hypothesis for the homologies of the digits of modern birds is I-II-III. This hypothesis implies that several phalanges have been lost during the evolution of modern birds (see Fig. 2).

Further evidence comes from the analysis of the trend of digital reduction in the outgroups of birds: various theropod dinosaurs. Most authors have suggested that theropod dinosaurs tend to lose or reduce the postaxial digits of the manus (digits IV and V, Figs. 2 and 3). The loss of metacarpal V is a synapomorphy of Theropoda (including Aves) and digit IV is reduced in several genera independently (Fig. 3 and Gauthier, 1986). In conjunction with the evidence from phalangeal formulae, this synapomorphy implies that the digits of *Archaeopteryx* are I-II-III and that, by extension, modern birds only possess I-II-III as well (Taquet, 1977).

Steiner (1922) used developmental evidence to suggest that the digits of the avian wing are homologous to digits I-II-III of other archosaurs. Steiner adhered to a strict view of Haeckelian recapitulation, one that led him to propose that five digits are present during the early development of the wing skeleton and that two of these digits regress during later development. Steiner's interpretations differ strongly from those of other embryologists in that Steiner's digit V represents the pisiform bone of other authors. Different recapitulationist interpretations have led other embryologists to support a II-III-IV view, as discussed below.

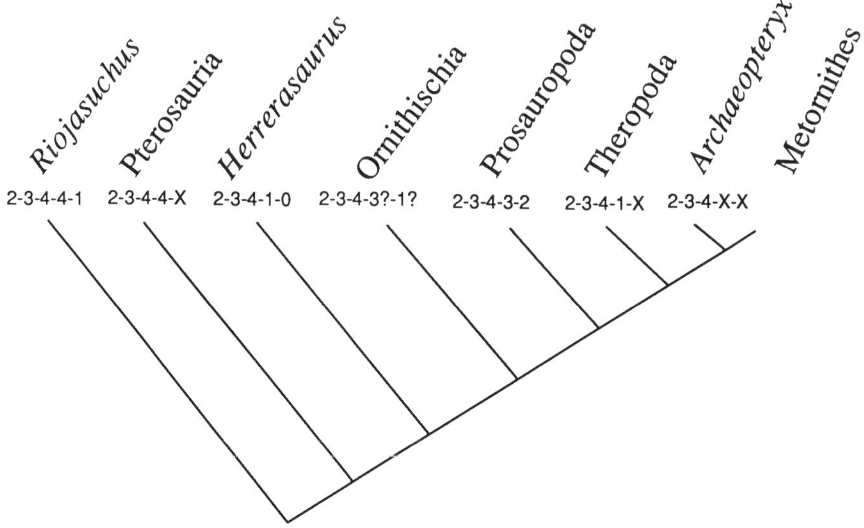

Figure 3. Phylogenetic relationships (after Gauthier, 1986; Sereno, 1991; and Altangerel et al., 1993) of selected archosaurs are included with the generalized phalangeal formulae of the terminal taxa. The taxon, Aviale, is a monophyletic group that includes *Archaeopteryx* and Metornithes (birds other than *Archaeopteryx*).

II-III-IV Theories

The II-III-IV interpretation of the digits of the avian manus has been supported by both embryological and paleontological data (Owen, 1836; Morse, 1872; Leighton, 1894; Prein, 1914; Lutz, 1942; Montagna, 1945; Holmgren, 1955; Tarsitano and Hecht, 1980; Thulborn and Hamley, 1982; Hinchliffe and Hecht, 1984; Shubin and Alberch, 1986; Müller and Alberch, 1989; Müller, 1991). Several criteria have been cited as evidence: (1) comparative analysis of the vascular supply of the avian manus; (2) recapitulation; (3) topological comparisons of precartilaginous blastema of chicks and other amniotes; (4) the sequence of chondrogenesis; and (5) the relative size of the digits of the manus.

Owen (1836) analyzed the vascular supply to the manus to conclude that digital reduction occurred on the anterior (radial) side of the limb and that, as a consequence, digit one has been lost. Owen noted that the radial artery is lacking in the birds that he dissected. Because this structure supplies the anterior digit in pentadactyl tetrapods, he proposed that the anteriormost digit (digit I) has been lost in birds.

Under the guise of recapitulation, some embryologists identified a limb archetype that made a transient appearance during the early development of the hand and foot. Holmgren (1955) proposed that all five metacarpals appear during the development of the avian wing. Holmgren observed that the elements he identified as digits I and V disappear at later stages. Most embryologists have been able to observe only four digits so the identifications rely on other evidence.

Montagna (1945) proposed that digits II-III-IV-V appear during avian wing development. These identifications relied on the fact that most other amniotes have a tendency to lose digit I. Morse's (1872) "Law of Digital Reduction" served as the basis for Montagna's (1945) view. Morse noted that the first digits to be lost in most amniotes are digits I and V. These outer digits have been reduced or lost many times independently in squamates and mammals. This dramatic evolutionary trend was extended to the analysis of the avian wing (Montagna, 1945; Hinchliffe and Hecht, 1984).

Hinchliffe and Hecht (1984) support the II-III-IV interpretation because of the presence of anterior and posterior necrotic zones in developing chick limbs. These anterior and posterior necrotic zones would tend to remove digits I and V. Hinchliffe and Hecht also suggest that the extreme postaxial border of the manus is defined by the position of the pisiform; consequently, digit IV cannot lie along this axis. And they identify a small postaxial condensation as a rudiment of digit V.

Shubin and Alberch (1986), Müller (1991), and Müller and Alberch (1989) use a strict interpretation of the developmental sequence to support a II-III-IV view. In most amniotes and anurans the first digit to form is digit IV, and this digit develops along a "primary axis" that consists of the continuous mesenchymal condensations that extend from the ulnare. In birds a similar axis develops. If the digit that develops along this axis is identified as digit IV, then the avian digits are II-III-IV.

Thulborn and Hamley (1982) suggest that the three digits of the avian wing, *Archaeopteryx*, and theropod dinosaurs are all homologous to II-III-IV of other tetrapods. The evidence for this view is that the axial digit (III) is considered to be "dominant" in five-digited tetrapods (Thulborn and Hamley, 1982, p. 6171). Thulborn and Hamley suggest that digit III is "dominant" in the sense that it possesses the largest metacarpal in other amniotes and it persists when other digits are lost. If the largest digit of the manus of birds, *Archaeopteryx*, and theropods is identified as homologous to digit III of other tetrapods, then the digital homologies are II-III-IV.

Tarsitano and Hecht (1980) reinterpret the existing material of *Archaeopteryx* to propose a new phalangeal formula for the manus. They extrapolate from this evidence to conclude that *Archaeopteryx* and modern birds possess digits II-III-IV, while theropods retain digits I-II-III. Tarsitano and Hecht cite the embryological evidence as support for their hypothesis.

Implications of Digital Homologies

Phylogenetic evidence (Fig. 3; Gauthier, 1986) indicates that (1) extant birds and *Archaeopteryx* are members of the a monophyletic group, Avialae, and that (2) the outgroups of Aviale are various non-avian theropod dinosaurs. This phylogenetic hypothesis is supported by numerous skeletal characters. No matter whether one accepts the I-II-III or the II-III-IV homology schemes, the distribution of characters that supports this hypothesis must be taken into consideration.

The phylogenetic scheme above suggests that the loss of metacarpal V and the reduction of metacarpal IV are derived characters of theropods. The identification of this synapomorphy is based on the conserved phalangeal formulae (2-3-4-x-x) in theropods and their outgroups, and the morphology of the anterior three digits in theropods. Derived theropods have only three digits and these three digits mostly retain a phalangeal formula of 2-3-4-x-x. These patterns strongly support the notion that theropods possess digits I, II, and III.

If one accepts the I-II-III hypothesis of homology, homoplasy of developmental sequence must be accepted. In most amniotes, digit IV is the first digit to develop. When it first appears, digit IV is connected to the precartilaginous primordia of the ulna, ulnare and other more proximal elements. These developmental features are known collectively as the "primary axis". Modern birds also possess a distinct primary axis. If one accepts the I-II-III scheme, then the metacarpal that forms within the avian primary axis is not digit IV (as in other amniotes) but is digit III. If this hypothesis is correct, digit III of birds is developmentally accelerated as compared to the other digits of the manus.

Gauthier's (1986) analysis reveals that theropods exhibit a very unusual sequence of digital reduction. Digital reduction in most amniotes (in both the manus and pes) follows a stereotyped pattern such that the first digits to be lost are digits I and V (Hinchliffe, 1991). In extreme cases (such as in those of the hands of lizards of the genus, *Lerista* [Greer, 1991]), digits II and then III will be lost. The apparent stability of digit I in the hand of theropods is unusual despite the fact that the foot shows the standard sequence of digital reduction. The phylogenetic interpretation of Gauthier (1986) unambiguously supports this view of non-avian theropod digital reduction. The application of this phylogeny to bird character evolution leads to homoplasy in developmental patterns (utilizing the I-II-III interpretation of avian digital homologies) or homoplasy in numerous other skeletal characters (in the II-III-IV interpretation of digital homologies).

There are several potential interpretations that result from accepting the II-III-IV hypothesis of digital homologies. All of these interpretations would weight developmental characters heavily and would lead to large amounts of homoplasy of other skeletal characters.

One could use the II-III-IV interpretation to specify a new phylogenetic hierarchy. This identification could be used to hypothesize that birds (including *Archaeopteryx*) are not related to theropod dinosaurs and are, in fact, related to more primitive archosaurs. To propose this view, one would have to accept a very large amount of homoplasy of characters of the skull, axial skeleton, and limbs. With regard to limbs, the phalangeal formulae and digital proportions of the manus of *Archaeopteryx* would be seen to evolve a similar pattern to theropods in parallel. At present, there is no compelling evidence to suggest this extreme homoplasy.

The II-III-IV hypothesis of homology could also be interpreted in light of the phylogenetic hierarchy proposed by Gauthier (1986). All of the above interpretations involve the homoplastic loss of digits, re-appearance of digit IV, alterations of phalangeal formulae, and (in some cases) convergence of the size and shape of the digits. One could accept that the digits of theropods are I-II-III and that the digits of birds (including *Archaeopteryx*) are II-III-IV. This scheme would involve (1) loss of digit I, (2) phylogenetic reappearance of metacarpal IV and four of its phalanges, (3) loss of two phalanges on digit II, and (4) addition of one phalange on digit III. Furthermore, one would have to accept that digits II-III-IV of *Archaeopteryx* convergently acquire the same proportions, size and shape as digits I-II-III of non-avian theropods. Alternatively, one could accept that the digits of *Archaeopteryx* and non-avian theropods are I-II-III and that extant birds share digits II-III-IV. This case involves a similar re-appearance of digit IV as well as the loss of phalanges on several of the interior digits. If one used the developmental information to re-interpret the digits of the primitive theropod hand as being II-III-IV-V, similar types of homoplasy would be encountered.

The I-II-III interpretation of the digits of the avian wing involves significantly less homoplasy than the II-III-IV hypothesis. The strongest arguments against the I-II-III interpretation involve (1) the interpretation of the primary axis, and (2) the unique sequence of digital reduction implied by the I-II-III scheme. The former entails a heterochrony in the formation of digit three. The latter difficulty is not affected by the analysis of avian digital homologies. The unique trend in the loss of digits four and five of theropods would be present regardless of whether one identifies the digits of birds as being I-II-III or II-III-IV. This example underscores the importance of fossils in the interpretation of developmental homologies. The interpretation of developmental patterns of birds must, perforce, be strongly affected by paleontological data. Without fossil theropods, and the phylogenetic interpretations that they imply, the II-III-IV interpretation of the homologies of the avian digits would, perhaps, be a more likely interpretation.

At first glance, the avian digits appear to be an example where biological and phylogenetic approaches to homology differ. In reality, this may not be the case. Recall that a hypothesis of biological homology is a statement about developmental constraints and relates to the manner in which characters can vary. The analysis above does not discuss the genetic and epigenetic controls of limb formation. Further study of the genetic and epigenetic controls of limb development may reveal that the wings of birds, and perhaps even the digits, are biologically homologous to those of other amniotes. Recent data on the genetic controls of limb development lends provisional support for this notion (Morgan et al., 1992; Tabin, 1991). Mouse and chick limbs share similar patterns of expression of Homeobox-containing genes, particularly those of the Hox-4 cluster. These genes are implicated in the development of a wide variety of structures and conserved patterns of expression indicate that the genetic controls of pattern formation in these limbs may be similar.

PRE-PHYLOGENETIC AND POST-PHYLOGENETIC USES OF DEVELOPMENTAL DATA

Developmental data has both pre- and post-phylogenetic uses in the analysis of homology. The pre-phylogenetic role of developmental data lies in the identification of characters for use in phylogenetic analysis. The post-phylogenetic application of developmental data provides *tests* of hypotheses of homology.

Developmental mechanisms provide information about the correlations between characters and can be used to identify individual characters for use in phylogenetic analysis. Genetic and epigenetic interactions circumscribe the limits to morphological integration and variation. An understanding of the structural basis of morphological integration enables a biologically meaningful atomism of features into discrete characters. Developmental studies of morphological integration involve the analysis of teratology, experimental manipulations of developmental processes, perturbations of the genetic factors involved with pattern formation, and the analysis of intra-specific variation. Each of these approaches enables the discrimination of potential developmental characters or homologies. These developmental criteria allow the formulation of hypotheses of homology that can be tested by phylogenetic analysis.

The phylogenetic analysis of developmental mechanisms is a vital component in the analysis of form because both developmental and phylogenetic information are essential in the causal analysis of homology. Historical, developmental analyses of form enable the identification of individual characters and can provide hypotheses about likely and unlikely modes of morphological transformation. Historical data provide not only tests of these hypotheses but also information about the manner in which developmental systems have, themselves, evolved.

Acknowledgments

I would like to thank Kevin de Queiroz, Kevin Padian, Olivier Rieppel, Paul Sereno, and Scott Winters for numerous discussions on these topics. This research was supported by NSF BSR 90068000.

References

Abel, O. 1912. *Grundzüge der Paleobiologie der Wirbeltiere*. Stuttgart: E. Schweizerbart.

Altangerel, P., M. A. Norell, L. M. Chiappe and J. M. Clark. 1993. Flightless birds from the Cretaceous of Mongolia. *Nature*

(London), 362: 626-633.

Bock, W. 1977. Foundations and methods of evolutionary classification. In M. K. Hecht, P. C. Goody and B. M. Hecht (Eds.), *Major Patterns of Vertebrate Evolution*, 851-895. New York: Plenum Press.

Broom, R. 1906. On the early development of the appendicular skeleton of the ostrich. *Transactions of the South African Philosophical Society*, 16: 355-369.

de Pinna, M. 1991. Concepts and tests of homology in the cladistic paradigm. *Cladistics*, 7: 367-394.

Eldredge N., and J. Cracraft. 1980. *Phylogenetic Patterns and the Evolutionary Process. Method and Theory in Comparative Biology*. New York: Columbia University Press.

Gauthier, J. 1986. Saurischian monophyly and the origin of birds. *Memoirs of the California Academy of Sciences*, 8: 1-55.

Gegenbaur, H. 1864. *Untersuchungen zur vergleichenden Anatomie der Wirbeltiere. I. Carpus und Tarsus*. Leipzig: W. Engelmann.

Goodwin, B. C., and C. Trainor. 1983. The ontogeny and phylogeny of the pentadactyl limb. In B. Goodwin, N. Holder, and C. Wylie (Eds.), *Development and Evolution*, 75-98. Cambridge: Cambridge University Press.

Greer, A. E. 1991. Limb reduction in the scincid lizard genus, *Lerista*. Variation in the bone complements of the front and rear limbs and the number of post-sacral vertebrae. *Journal of Herpetology*, 24: 142-150.

Haszprunar, G. 1992. The types of homology and their significance for evolutionary biology and phylogenetics. *Journal of Evolutionary Biology*, 5: 13-24.

Heilmann, G. 1927. *Origin of Birds*. London: Witherby.

Hinchliffe, J. R. 1977. The chondrogenic pattern in chick limb morphogenesis: a problem of development and evolution. In D. Ede, J. R. Hinchliffe and M. Balls (Eds.), *Vertebrate Limb and Somite Morphogenesis*, 293-309. Cambridge: Cambridge University Press.

Hinchliffe, J. R. 1989. Reconstructing the archetype: innovation and conservatism in the evolution and development of the pentadactyl limb. In D. Wake and G. Roth (Eds.), *Complex Organismal Functions: Integration and Evolution in Vertebrates*, 171-188. London: John Wiley and Sons.

Hinchliffe, J. R. 1991. Developmental approaches to the problem of transformation of limb structure in evolution. In J. Hinchliffe, J. Hurle and D. Summerbell (Eds.), *Developmental Patterning of the Vertebrate Limb*, 313-323. New York: Plenum Press.

Hinchliffe, J. R., and P. Griffiths. 1983. The pre-chondrogenic patterns in tetrapod limb development and their phylogenetic significance. In B. Goodwin, N. Holder and C. Wylie (Eds.), *Development and Evolution*, 99-121. Cambridge: Cambridge University Press.

Hinchliffe, J. R., and M. Hecht. 1984. Homology of the bird wing skeleton. *Evolutionary Biology*, 20: 21-37.

Hinchliffe, J. R., and D. R. Johnson. 1980. *The Development of the Vertebrate Limb*. London: Oxford University Press.

Holmgren, N. 1955. Studies on the phylogeny of birds. *Acta Zoologica, Stockholm*, 36: 243-328.

Huxley, T. 1868. On the animals which are most nearly intermediate between birds and reptiles. *Annals and Magazine of Natural History*, 4: 2.

Jeffries, J. A. 1881. On the fingers of birds. *Bulletin, Nuttall Ornithology Club*, 6: 6-11.

Leighton, A. 1894. The development of the wing of *Sterna wilsoni*. *American Naturalist*, 28: 678-774.

Lutz, H. 1942. Beiträge zur Stammesgeschichte der Ratiten. Vergleich zwischen Emu-Embryo und entsprechendem Carinatenstadium. Diss. Genève. (cited in Holmgren, 1955)

Meckel, J. F. 1825. *System der vergleichenden Anatomie*. Vol. 2. Halle: Rengersche Buchhandlung.

Minelli, A., and B. Peruffo. 1991. Developmental pathways, homology, and homonomy in metameric animals. *Journal of Evolutionary Biology*, 3: 429-445.

Montagna, W. 1945. A re-investigation of the development of the wing of the bird. *Journal of Morphology*, 76: 87-118.

Morgan, B., J. C. Izpisua-Belmonte, D. Duboule and C. Tabin. 1992. Targeted misexpression of Hox-4.6 in the avian limb bud causes apparent homeotic transformations. *Nature (London)*, 358: 236-239.

Morse, E. 1872. On the carpus and tarsus of birds. *Ann. Lyc. Natural History*, New York, 10: 3-22.

Müller, G. 1990. Developmental mechanisms at the origin of morphological novelty: a side effect hypothesis. In M. Nitecki (Ed.), *Evolutionary Innovations*, 99-130. Chicago: The University of Chicago Press.

Müller, G. 1991. Evolutionary transformations of limb pattern: heterochrony and secondary fusion. In J. Hinchliffe, J. Hurle and D. Summerbell (Eds.), *Developmental Patterning of the Vertebrate Limb*, 395-405. New York: Plenum Press.

Müller, G., and P. Alberch. 1989. Ontogeny of the limb skeleton in *Alligator mississippiensis*: developmental invariance and change in the evolution of archosaur limbs. *Journal of Morphology*, 203: 151-164.

Müller, G., and G. Wagner. 1991. Novelty in evolution: restructuring the concept. *Annual Review of Ecology and Systematics*, 22: 229-256.

Nelson, G. 1978. Ontogeny, phylogeny, paleontology and the biogenetic law. *Systematic Zoology*, 27: 324-345.

Ostrom, J. 1976. *Archaeopteryx* and the origin of birds. *Biological Journal of the Linnean Society*, 8: 91-182.

Owen, R. 1836. Aves, in *Todd's Cyclopedia of Anatomy and Physiology*, Vol. 1, pp. 265-358.

Padian, K. 1992. A proposal to standardize tetrapod phalangeal formula designations. *Journal of Vertebrate Paleontology*, 12: 260-262.

Parker, W. K. 1888. On the structure and development of the wing in the common fowl. *Philosophical Transactions of the Royal Society of London B*, 179: 385-395.

Patterson, C. 1982. Morphological characters and homology. In K.

Joysey and A. E. Friday (Eds.), *Problems of Phylogenetic Reconstruction*, 21-74. London: Academic Press.

Patterson, C. 1988. Homology in classical and molecular biology. *Molecular Biology in Evolution*, 5: 603-625.

Prein, F. 1914. Die Entwicklung des vorderen Extremitätenskelettes beim Haushuhn. *Anatomie*, 51: 643-690.

Rieppel, O. 1980. Homology, a deductive concept? *Zeitschrift für Zoologische Systematik und Evolutionsforschung*, 18: 315-319.

Rieppel, O. 1988. *Fundamentals of Comparative Biology*. Basel: Birkhäuser.

Romer, A. 1956. *Osteology of the Reptiles*. Chicago: University of Chicago Press.

Rosenberg, A. 1873. Über die Entwicklung des vorderen Extremitätenskelettes bei einigen durch Reduktion ihrer Gliedmassen charakterisierten Wirbeltiere. *Zeitschrift für Zoologische Wissenschaftliche*, 23: 116-166.

Roth, V. L. 1984. On homology. *Biological Journal of the Linnean Society*, 22: 13-29.

Roth, V. L. 1988. The biological basis of homology. In C. J. Humphries (Ed.), *Ontogeny and Systematics*, 13-29. New York: Columbia University Press.

Roth, V. L. 1991. Homology and hierarchies: problems solved and unresolved. *Journal of Evolutionary Biology*, 4: 167-194.

Sereno, P. 1991. Basal archosaurs: phylogenetic relationships and functional implications. *Journal of Vertebrate Paleontology*, 11 (suppl. 4): 1-53.

Shubin, N. 1991. The implications of "The Bauplan" for the development and evolution of the tetrapod limb. In J. Hinchliffe, J. Hurle and D. Summerbell (Eds.), *Developmental Patterning of the Vertebrate Limb*, 411-421. New York: Plenum Press.

Shubin, N., and P. Alberch. 1986. A morphogenetic approach to the origin and basic organization of the tetrapod limb. *Evolutionary Biology*, 20: 319-387.

Siegelbauer, F. 1911. Zur Entwicklung der Vogelextremität. *Zeitschrift für Zoologische Wissenschaftliche*, 97: 262-313.

Steiner, H. 1922. Die ontogenetische und phylogenetische Entwicklung des Vogel-Flugelskelettes. *Acta Zoologica, Stockholm*, 3: 1-54.
Tabin, C. 1991. Reninoids, homeoboxes, and growth factors: toward molecular models for limb development. *Cell*, 66: 199-217.
Taquet, P. 1977. Variation ou rudimentation du membre anterieur chez les Theropodes (Dinosauria)? *Colloques internationaux. Centre national de la Recherche Scientifique*, 266: 333-339.
Tarsitano, S., and M. Hecht. 1980. A reconsideration of the reptilian relationships of *Archaeopteryx*. *Zoological Journal of the Linnean Society*, 69: 149-182.
Thulborn, R., and T. Hamley. 1982. The reptilian relationships of *Archaeopteryx*. *Australian Journal of Zoology*, 23: 611-634.
Van Valen, L. 1982. Homology and causes. *Journal of Morphology*, 173: 305-312.
Wagner, G. P. 1989a. The origin of morphological characters and the biological meaning of homology. *Evolution*, 43: 1157-1171.
Wagner, G. P. 1989b. The biological homology concept. *Annual Review of Ecology and Systematics*, 20: 51-69.
Zehntner, L. 1890. Beiträge zur Entwicklung von *Cypselus melba*. *Archiv für Naturgeschichte*, 56: 189-220.

9

Summary and Comments on Systematic Pattern and Evolutionary Process

Olivier Rieppel and Lance Grande

Department of Geology
Field Museum of Natural History
Chicago, Illinois 60605

SOME OF THE MOST FUNDAMENTAL ISSUES that scientists and philosophers face in evolutionary studies today concern the relation of systematic patterns observed in nature to evolutionary process theories proposed to explain them (including metatheories *sensu* Nelson, 1989, such as Darwinism, neodarwinism, punctuationism, structuralism). We could ask, what are the connections between phylogenetic patterns and evolutionary process theories, and how far have we come in the scientific validation and development of these connections? Hierarchical patterns of taxa are descriptive summaries, and the observed data from organisms used to generate these patterns are discontinuous sets (i.e., taxa). Although an evolutionary continuum between taxa can be theorized, all such continuity is theoretical rather than observable. We do not directly observe macroevolutionary processes such as speciation, cladogenesis, or anagenesis. We propose them as mechanisms for theoretical connections. Phylogenetic patterns (i.e., cladograms) are parsimonious or maximally compatible summaries of character sets belonging to discrete taxa. In some sense, discontinuities allow us to differentially diagnose and identify terminal taxa, which we then place in a theoretical continuum that assumes all life to be related (descended from a common ancestor). Evolutionary process theory connects all living organisms to each other historically.

 Copyright © 1994 by Academic Press, Inc.
All rights of reproduction in any form reserved.

Some of the processes used to explain evolution are both observable and continuous on a relatively fine scale. These processes of origination and change (e.g., birth, certain aspects of development, and parent-offspring descent lineages) can be observed by study and description of living organisms. Many or most evolutionary scientists have extrapolated from studies of observably continuous phenomena such as these to theorize a continuum (evolutionary connection) between discrete taxa and explain patterns of internested similarities. They infer elements of analogy between birth and speciation, human genealogy and phylogeny, and death and extinction. Indeed, as noted by Gilmour (1961, p. 35), much of the most basic terminology of macroevolutionary process was taken from the study of human genealogy (e.g., "ancestor", "descendant", phylogenetic "tree", relative "relationship"). So, perhaps another question that evolutionary scientists should ask today is, "How far have we come in validating or discarding the extrapolation of observed process studies to explain phylogenetic patterns?" This question relates to previous notions of *micro-* versus *macroevolutionary* processes. Simpson (1944, p. 97), for example, defined these two terms as follows:

> "Micro-evolution involves changes within potentially continuous populations, and there is little doubt that its materials are those revealed by genetic experimentation. Macro-evolution involves the rise and divergence of discontinuous groups (taxa), and it is still debatable whether it differs in kind or only in degree from micro-evolution."

Microevolutionary process is sometimes observable, while macroevolutionary process (e.g., speciation) is never definitively observable, but is instead an extrapolation of microevolutionary processes to explain phylogenetic patterns. Proponents of the "synthetic theory of organic evolution" (e.g., Dobzhansky et al., 1977) frequently have tried to shrink the theoretical leap required to bridge from microevolutionary data to macroevolutionary theories. For example, Dobzhansky (1970, p. 429) stated:

> "A loose distinction is often drawn between microevolution and macroevolution, according to the magnitude of the morphological, physiological and genetic changes. One may say that microevolutionary changes are possibly reversible, whereas macroevolutionary ones are not. This does not mean that there are two kinds of evolution, micro- and macro-, as was contended by some authors.... The distinction is quantitative and any boundary can only be arbitrary."

We do not see the boundary between micro- and macroevolutionary process theory as merely quantitative. We see it as largely qualitative and methodological. Microevolutionary process theories are appealing, and some of them are necessary (in a broad sense) but not sufficient to explain macroevolutionary patterns. Microevolutionary studies can use descriptive data to make predictions within a continuous lineage (e.g., Mendelian rules of inheritance; Hardy-Weinberg Principle for population genetics). But macroevolutionary data (phylogenetic patterns or cladograms) are so far not known to allow analogous predictions. We do not observe macroevolutionary process (or at least we have no way of predicting noncyclic change in nature in the form of speciation or anagenesis and subsequently testing that prediction). The observation of microevolutionary process is done in real (observable) time on living organisms. The connection of those processes to phylogenetic pattern as macroevolutionary process is still a theory of extrapolation.

If we are unable to observe and definitively identify macroevolutionary processes, what could we do to tie macroevolutionary process theories more closely to pattern? Maybe nothing. Or maybe we have yet to discover new causal links between different types of matching patterns that would bind pattern and process more closely. Evolutionary science may still await its most major breakthrough.

The distinction between systematic pattern and evolutionary process became particularly prominent with the acceptance of Hennig's (1966) principles of phylogeny reconstruction (Eldredge and Cracraft, 1980; Nelson and Platnick, 1981; Wiley, 1981). Analysis of character distribution allows discovery of noncausal (i.e., ahistorical) hierarchical patterns in nature to be expressed in a cladogram (branching diagram). Evolutionary process theory is a causal (i.e., historical) explanation for these patterns.

Later the community of "Hennigians" or "cladists" were classified as belonging to two separate camps, the so-called "phylogenetic systematists" versus the "transformed" or "pattern cladists" (Beatty, 1982). The first school roots the search for pattern in the axiomatic assumption of evolution (de Queiroz, 1988). That assumption permits the deduction that systematic pattern is a direct reflection of the course of phylogeny over time. The cladogram is viewed in a temporal context as a phylogenetic tree and the branching points as speciation events (e.g., Wiley et al., 1991; Brooks and McLennan, 1991). The opposing school, that of "pattern cladists", views the cladogram as a time-independent summary of character distribution as observed today rather than an evolutionary tree (Platnick, 1977; Patterson, 1980; Brady, 1985; Rieppel, 1988). The time dimension, and with it the theory of evolution, are added to the cladogram in an effort to explain congruent character distribution (i.e., hierarchy). Cladograms are understood in terms of the relative relationship of terminal taxa, rather than in terms of species origin through branching events.

In the second chapter of this book, Brady addresses the controversy between pattern cladists and phylogenetic systematists, arguing the pattern cladists' view that the search for pattern must proceed independently from explanatory theories. His starting point is in agreement with the philosophical thesis that there can be no theory-free observation. But he then makes the point that if one allows the theory used to explain certain phenomena to influence the search for those phenomena, one is likely to obtain the results one is looking for. Philosophers call such a procedure circular, and the theory empirically empty, because it can never be shown to be wrong with respect to observational data (Brady, 1982). Pattern cladists contend that observational data should primarily be descriptive and contaminated with as few preconceived explanatory theories as possible. From the perspective of a historian of evolutionary science, Brady shows that comparative morphology (i.e., description of form) predated Darwinism, and discovered relations, or "affinities", among organisms, without Darwin's theory of evolution through natural selection. Darwin (1859) maintained that "affinities" discovered by his predecessors should be taken as evidence of "family relationships", which he explained by common descent.

If it was possible before Darwin to pursue a descriptive program in comparative biology independent of evolutionary theory, then it should still be possible to do so today. The search for pattern cannot be entirely theory free, but it can be independent from evolutionary theory. Theories influencing the search for pattern would still be dependent on a theory of hierarchy, a theory of characters used for comparison, a theory of similarity, and so on. But the theory of evolution can be added to the pattern obtained as its causal explanation, and hence would be saved from empirical emptiness.

A number of objections have been raised against the pattern cladists' program. Ridley (1986), for example, contends that dissociating the search for pattern in nature from evolutionary theory would rob the search of its theoretical justification. Why, he asks, should we look for order in nature if not to learn something about evolution? Others deny the possibility of thinking in a pre-Darwinian context, searching for systematic patterns as if evolutionary theory had never been proposed (i.e., how, they ask, can any systematic biologist in this day and age be sure to block all Darwinian bias?). And if the branching order expressed in a cladogram indicates diversification of natural groups over time, they ask, what else could the nodes in the cladogram mean other than speciation? After all, evolution is about the origin of new species, as Darwin implied in the title of his book. New species originate through speciation, and speciation is the only process we know that causes diversification of natural groups.

But how do we know? is the pattern cladist's (and our) main response. As Patterson (1980, p. 239) asks us, "...how can the process of evolution be discerned except through the pattern of character distribution?" How do we know what taxa are, how new species originate, and how natural groups diversify through speciation? All we observe are organisms, their geographic occurrence, their characters (morphological, molecular, and other), and the change those characters undergo during individual development (ontogeny). All other processes, such as speciation, diversification of natural groups, and their changing geographic distribution through geological time, must be inferred from those observable data. Although characters and taxa are thought to have been caused by descent, taxa are discovered through characters that are themselves discovered through empirical investigation (Nelson and Patterson, 1993).

Consider species. We can think of species as representing a category in Linnean classification, in which case it is a mere convention, a way we find convenient to classify organisms. We can also think of species as particular groups of organisms, such as ostriches, eastern gray squirrels, or bluegill sunfishes. Species as groups of organisms are taxa, but a mere enumeration of taxa has little meaning. Only if species as taxa have something in common with each other does the species concept have theoretical meaning, because only then is it possible to make general statements about species that would represent a scientific theory of phylogeny.

Species have been defined as groups of interbreeding organisms, reproductively isolated from other such groups of organisms (Mayr, 1942, 1963). Speciation is generally viewed as a process whereby an ancestral species is split, by whichever historical cause, into two or more sister species that eventually acquire reproductive isolation (Otte and Endler, 1989). Such a model of diversification of taxa can be construed as an exact fit to the phylogenetic systematist's view of the cladogram.

The problem as we see it, however, is that speciation transcends observable time. Sister species are thought to have evolved from exclusive common ancestral species represented by nodes on the cladogram which, in a very general sense, identify hypothetical ancestors. Speciation, like common ancestry, cannot be observed, but must be inferred as an explanation for shared similarity, i.e., homologies shared by the two sister taxa. Natural groups (taxa) are diagnosed on the basis of homology. Whatever species are claimed to be, our knowledge about species diversity and interrelationships seems to come from pattern reconstruction based on the relation of homology. Looking at the "species problem" from this perspective creates a difficulty for process explanations. The Darwinian theory of evolution through variation and natural selection hypothesizes that new species result from a gradual and step-wise process in the genealogical continuum of successive generations. Discrete character recognition would become blurred in a continuous process of gradual change, which is why Darwin claimed that species are mere conventions, not really existing in nature, and that the perceived gaps between species are due to extinction rather than saltationary change (Mayr, 1982). One logical yet untestable alternative to gradual evolution is to postulate discontinuous species change, as has been done by proponents of macromutations (Goldschmidt, 1940), quantum evolution (Simpson, 1944), and punctuated equilibrium (Eldredge, 1971; Eldredge and Gould, 1972; Gould and Eldredge, 1977). Another alternative is to claim that species are not diagnosed by homologies. This results in the distinction of two categories of putatively natural taxa: species (non-monophyletic by this definition), and monophyletic groups of species. Proponents of this view include Wiley et al. (1991) and Brooks and McLennan (1991). According to these authors, only monophyletic groups would be diagnosed on the basis of homology; the species would correspond to a reproductively isolated population, irrespective of the shared similarity. But if that is so, how do we learn about species diversity and interrelationships? The argument has become caught in a circle. It seems to us that we cannot have it both ways at the same time. We cannot use homologies and a theory of relations among taxa to investigate species and the process of speciation and at the same time use models of speciation to give us a clue to systematic patterns and homology identification. A unifying theoretical framework for evolutionary biology may be a scientific ideal–yet impossible to achieve. This, indeed, is the position argued by Beatty in the third chapter of this book, adopting the premise that theoretical pluralism prevails.

Beatty starts with the Newtonian ideal of science (i.e., the ideal to discover universal truth through scientific methods). Newton held an outstanding position among his fellow scientists of the 17th and 18th centuries because he, for the first time, developed a scientific method that hinted at a unifying theory of the earth. His law of gravity explains the relative motion of bodies with respect to one another without the need for these bodies to come into physical contact. With respect to the medieval "impetus" theory, this seemed like taking recourse to an "occult" force. Yet Newton's laws of physics were quickly taken into biology to explain the orderly formation of an embryo from parts contributed by both parental bodies. Here originated the desire of biology to live up to the theoretical and practical stringency of physics, an agenda that Darwin also declared to be his own. Darwin initially felt that his Law of Natural Selection deserved equal status to Newton's Law of Gravity (Ospovat, 1981), i.e., the status of a time-independent law (secondary cause) enacted by the Creator and revealing a universal truth guiding the natural course of events.

Beatty's alternative proposition to the Newtonian ideal is theoretical pluralism. We can illustrate the notion of theoretical pluralism by looking at the nature of light. We all think we know what light is, yet classical physics has been unable to provide us with a unifying account of this phenomenon. Light can be understood in terms of waves, but it can also be understood in terms of moving particles. Both theories are relevant and testable with respect to some properties of light, yet no single theory can cover all aspects of light, and none covers the (unattainable) truth about light. In the last analysis, theoretical pluralism is a consequence of the impossibility of postulating a reality-in-itself waiting to be discovered by scientists, whose methods of discovery would enable them to make true statements about this reality.

Beatty's contrast between theoretical pluralism and the ideal of "unifying explanations" poses a problem, in our opinion. Although he appreciates the dependence of theoretical pluralism on purpose and goals (aims) of scientific investigation, he makes the statement that "'theoretical pluralism'...does not concern multiple *aims*; rather, it concerns multiple *theories*" (p. 36). By that statement, he implies that there can be theories independent of aims (goals and purposes). However, as we see it, it is the scientists who ask questions, thus identifying phenomena worthy of investigation and amenable to hypothetical explanation. To paraphrase Popper (1963): Take a lecture-hall full of students and ask them to observe! They will necessarily have to ask the question, "What is it that we should observe?" The question of purpose and goal necessarily enters any and all scientific investigations at some level, and cannot be decoupled from them.

We contend that the attempt to decouple purpose and goal from scientific investigation results in a number of problems with Beatty's exploration of theoretical pluralism within the realm of biological systematics. "Methods", he writes, "should be judged...in terms of the *extent* of their applicability" (p. 41). The only way to judge the applicability of a scientific method is to ask what it was, or is, *intended* to discover, that is, what the purpose and goal of the investigation were, or are. Beatty goes on to assert that "insofar as these methods play a role in explaining patterns of similarity and dissimilarity among organisms..., systematics bears all the marks of theoretical pluralism..." (p. 41). That systematics should be subject to theoretical pluralism, as long as it is an empirical science, is hardly surprising (even if this is not acknowledged by some of its practitioners). But in our view, hypothetical explanations follow the discovery of regularity among phenomena, and methods are used to discover phenomena rather than explain them. Beatty states that "parsimony...plays an important role in explanations of character distributions" (p. 42). We fail to see how this could be. In our view, parsimony is a methodological correlate for the discovery of hierarchical patterns of character distribution (Brady, 1983), and its use (as a method) is justified not by reference to evolutionary explanations of natural hierarchies, but by the purpose of discovering the most complete hierarchy buried in any data set.

Our views also differ from Beatty's on the use of ontogeny as a method of discovering hierarchical patterns of character distribution. Following Beatty (p. 44), the statement that ontogeny proceeds from the more general to the less general condition of form does not apply in cases where more general characters arise in ontogeny from less general characters: "So, for instance, deletion of an ancestral character from the ontogeny of descendant taxa is ruled out if the generalization is really true" (p. 45). Knowledge of what is "really true" is irrelevant here because it is unknowable. De-differentiation of cells in an amputated urodele limb prior to its regeneration has been cited as an exception to the ontogenetic method (Kluge, 1985). But the ontogenetic method was not *intended* to address problems of regeneration. Instead, it was intended to serve as a discovery procedure for hierarchical character distribution, and in that context we fail to see how deletion of an ancestral character from the descendant ontogeny would falsify it as a method. This would be true only if the method was seen as part of the explanation, as it seems to be by Beatty. The presence of limbs follows the absence of limbs in ontogeny, and the presence of limbs (in tetrapods) is the less general condition compared to the absence of limbs (in fishes, sea urchins, flies, worms, plants, bacteria, etc.). If snakes are classified as a subgroup of tetrapods, the evolutionary *explanation* for the loss of limbs in snakes must be the deletion of an ancestral character in a descendant ontogeny. But that does not falsify the ontogenetic *method*, because ontogeny does not produce limbs in snakes which are then resorbed. Instead, limbs simply don't develop; ontogeny grinds to a halt at the more general condition of form! The ontogenetic method was not intended to discover, or explain, the evolutionary history of organisms. It was intended to discover a hierarchy of characters.

If nature were chaotic, if there were no order among living beings, natural scientists would have nothing to discover. Lack of regularity of phenomena, lack of order in nature, cannot be the focus of attention of systematic biology. We believe the interest of systematists is not attracted by chaotic character distribution, and no process explanations seem required to explain irrational chaos. It is order that systematists should look for, not disorder. Grande (Chapter 4, this volume) maintains that systematists search for repeated patterns in nature, because the more often patterns of similarity, or inferred relationship, repeat themselves in nature, the more likely it is that the patterns repeat some nonrandom phenomenon. Multiple repeated patterns in nature invite causal explanations and suggest the possibility of constant underlying causes in biology that are similar to those thought to prevail in physical sciences. The problem, however, is to determine what special information, if any, from repeated patterns might lead us to the discovery of such constants.

Grande (Chapter 4, this volume) evaluates independent patterns of relationship, among geographic areas, shown by different taxa. In that way, taxonomically independent patterns are matched against one another in a research program called "vicariance biogeography" (see also Nelson and Platnick, 1981; Grande, 1985; Humphries and Parenti, 1986; Wiley, 1988a, 1988b; Sober, 1988; Cracraft, 1988; Brooks, 1990). If most or all of the independent biogeographic patterns within a monophyletic group are congruent, the need for a unifying causal explanation exists. In a sense, this biogeographic method seeks homologous cladistic patterns rather than homologous characters. Homologous patterns lend support to specific theories of area relationships if they repeat enough times to be statistically significant (i.e., appear to be a nonrandom pattern in need of causal explanation).

In the search for repeating patterns, and in support of "taxonomic congruence" between different data sets, Grande suggests that we should take a step beyond area patterns and look for repeating patterns in studies of phylogeny. If homologous patterns can provide information relevant to the study of area relationships (vicariance biogeography), should they not also have information relevant to the study of taxonomic relationships (phylogeny)? Characters of organisms come in different kinds, as emphasized by Wake (Chapter 7, this volume). There are skeletal characters, myological characters, neuroanatomical characters, molecular characters, behavioral characters, and so on. Grande notes that different kinds of characters can also be analyzed independently in the search for systematic patterns. If congruence exists between patterns shown by different sets of data, this may indicate a strong historical signal.

Grande (Chapter 4, this volume) draws attention to the point that in practice, character systems are normally analyzed separately (by different specialists), and character equivalence may be impossible to demonstrate. For example, many morphological data of sperm discussed by Wake are based on single samples (M. Wake, pers. commun.), with little or no information on developmental or individual variation. We believe that addition of large amounts of such tenuous data to data from better known morphological systems could conceivably obscure a weak phylogenetic signal.

Various authors (Kluge, 1989; Rieppel, 1994) have voiced concern about subdividing the total data set known for a group of organisms. What, after all, is a character, and how does it contribute to the search for patterns of relationship? Identification of a character starts with the perception of some similarity, but the phylogenetic information content of shared similarity is determined by congruence of characters. It is not the character *per se* that conveys a phylogenetic signal, nor the number of characters, but the pattern of congruence among all characters known at any particular time. A pattern conveyed by a set of congruent characters is unlikely to be obscured by other data that are highly incongruent among themselves. The search for regularity of phenomena, that is, for congruence of character distribution, is the search for organisms sharing the greatest number of attributes in a congruent manner. Since there is no theory-free observation, we can never know, in an objective way, what a "shared character" is. A "shared character" can be only a hypothesis of homology to be tested by congruence. This is in agreement with the statement made above that the purpose of systematics is to discover order rather than chaos. The "test of congruence" (Patterson, 1982) is a means to find the maximally congruent (i.e., most regularly distributed) shared characters among the organisms under investigation. But if that is true, any one perceived character should be tested against all other characters known. Osteological characters should not be tested only against other osteological characters, but also against, for example, myological, neuroanatomical, and molecular characters. Only the "total evidence" method of including all characters in a single data set would provide a maximally rigorous test of congruence for an individual character.

Some aspects of the "principle of total evidence" are problematical, too. These originate with the combination of widely disparate characters in a unified analysis. Consider the comparison of morphological and molecular data. Similarity of morphology is usually perceived in qualitative terms, while similarity among molecules is more of a quantitative problem. How would the two sources of data ever be fully comparable? This is the issue addressed by Novacek (Chapter 5, this volume), who uses the controversial case of bat diphyly as an example. Incompatibility of data is only one side of the coin. The other is differential weighting of characters. If all known characters are used in conjunction, each contributes to the analysis with equal weight. If a large set of highly incongruent molecular data is combined with a smaller set of highly congruent morphological characters, congruence will convey the phylogenetic signal inherent in the combined data set. However, if a large molecular data set supports a hypothesis of relationships different from that of a small morphological data set, the phylogenetic evidence of the larger molecular data set can outweigh that of the smaller morphological data set. The sheer number of molecular characters may be much more influential to a phylogenetic analysis. If relative relationships are investigated on the basis of molecular evidence alone (as argued by Ho, 1988; quoted and further discussed by Wake, Chapter 7, this volume), then the implicit claim is that molecular evidence is more important, i.e., weighs more heavily in phylogenetic analysis than other kinds of characters such as morphological ones. If fossils are combined with extant organisms in one analysis, only skeletal characters can be used, or all nonskeletal characters must be coded as unknown for fossils. The implications of the principle of total evidence have nontrivial consequences for the combination of extant and fossil organisms in a single analysis, as is further discussed by Novacek. The same, however, is true for taxonomic congruence argued by Grande. If different kinds of characters are analyzed separately and yield congruent cladograms, then there is a strong signal for orderliness. If, however, different character sets yield incongruent hypotheses of relationships, then how can an objective choice be made as to which one is better supported? This is a particularly difficult issue if the quantity of data differs significantly, as is commonly the case in morphological versus molecular data sets. Again, reference to Beatty's notion of "theoretical pluralism", or perhaps more appropriately, a notion of "methodological pluralism", may be used as a warning against pitching total evidence versus taxonomic congruence as opposing methodologies in the search for patterns. It seems that the best avenue for pattern analysis is to continue to use both methods in conjunction, and to compare the results obtained in each case.

Morphological and molecular, as well as physiological, behavioral, and other different types of data, are used in identification of patterns that are not necessarily tied to history but can be explained through a historical process. The cladogram is not necessarily a temporal concept but, as mentioned above, the most efficient (congruent) summary of observed character distribution. Explanation of a cladogram by the historical process of evolution adds a time dimension to systematic pattern, indicating that yet another kind of data could be used in pattern identification–stratigraphic data. Phylogenetic explanation of the cladogram implies a relative ordering of taxa and branching events in time. Since taxa and branching events are identified by means of homologies, the cladogram also implies a relative temporal order of the emergence of evolutionary novelties.

Turning the argument around, Fisher (Chapter 6, this volume) contends that the stratigraphic order of the fossil record provides independent evidence to choose among conflicting phylogenetic patterns or even in some cases to overturn cladograms based on morphological characters. If morphological, molecular, or other characters yield several equally or nearly equally parsimonious cladograms, i.e., several hypotheses of relationship with closely comparable congruence within conflicting data sets, then, according to Fisher, stratigraphic information on the relative timing of appearance of the taxa in the fossil record may provide important additional information with which to choose among conflicting cladograms. Recent studies have indicated that in some cases there is a fairly close correlation between the relative order of branching events as indicated by cladograms, and the relative timing of the appearance of taxa in the stratigraphic column, at least for taxa well documented in the fossil record (Norell and Novacek, 1992a, 1992b). But other well-documented examples illustrate striking exceptions to the expectation that the oldest fossil of a clade will also show the greatest number of "primitive" character states (e.g., Evans and Hecht, 1993).

Using the hypothesis of congruence of stratigraphic data and phylogenetic patterns as a premise for the choice among several equally or nearly equally parsimonious cladograms, Fisher (1992, p. 125) introduced the novel concept of "stratigraphic parsimony debt", defined as "the number of independent discrepancies between expected and observed order of stratigraphic occurrence".

In our view, however, it is one thing to investigate, *a posteriori*, the congruence between phylogenetic patterns and stratigraphic occurrence (Norell and Novacek, 1992a, 1992b), but it is another thing to use "the number of independent discrepancies between expected and observed order of stratigraphic occurrence" to choose among competing phylogenetic hypotheses (Fisher, 1992, p. 125). Fisher makes the claim (Chapter 6, p. 157) that "If stratigraphic patterns are strong enough, they may overturn cladograms (and unrooted networks) supported by morphology alone"; and elsewhere (1992, p. 128), he states "There may be trees that are less than maximally parsimonious morphologically, but that make up for this by being more parsimonious stratigraphically". In that sense, Fisher uses stratigraphic information as data in phylogeny reconstruction, which, in his opinion, "represents a move in the direction of treating 'total evidence'" (Chapter 6, p. 165).

In our view there are two major challenges to the "stratocladistic method". First, the *strength* of a stratigraphic pattern cannot be measured independent of a pre-existing taxonomy (which is based almost exclusively on morphology). So, the stratigraphic order of fossils is not actually providing *independent* evidence to choose among conflicting cladograms. Second, we find the method's degree of dependence on the fossil record problematic (at least with regard to the method overturning cladograms based on morphology).

In general, it seems to us that many evolutionary biologists still underestimate the effects of the incompleteness of the fossil record, and are too dependent on negative evidence (inferring, for example, that the absence of fossil taxa from a stratum provides significant evolutionary data). Consider the hypothetical example illustrated in Figure 1, where the known stratigraphic distribution of three taxa (A, B, and C) is compared to two conflicting hypotheses of phylogenetic relationships

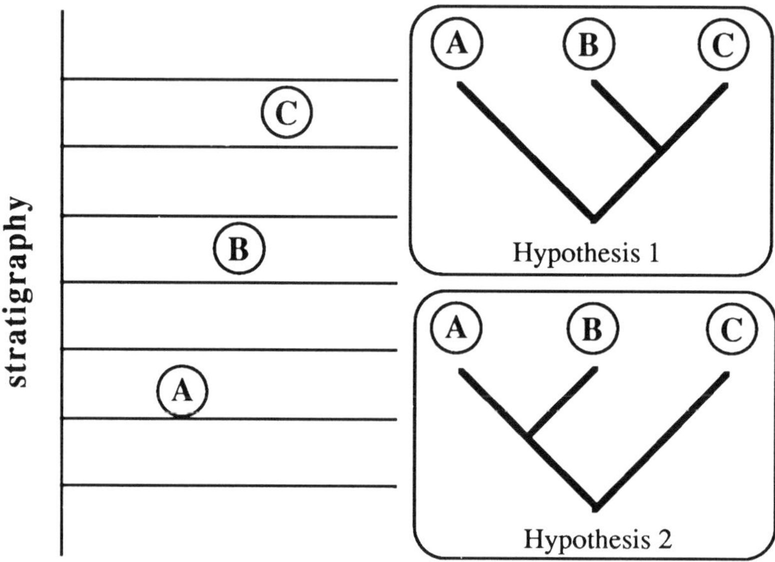

Figure 1. The known stratigraphic distribution of the taxa A, B, and C is compared to two competing cladograms.

based on character congruence. Hypothesis 1 would predict that B and C share a hypothetical common ancestor that evolved after the hypothetical common ancestor of A, B, and C. Hence it would be predicted that the branch including A would be represented in the fossil record before B or C, if stratigraphy were to reflect the phylogeny represented in hypothesis 1. Hypothesis 1 is in accordance with the known stratigraphic distribution of the taxa involved. Such is not the case for hypothesis 2, which would predict taxon C to be the first one to appear in the fossil record. Based on known stratigraphic distribution of taxa A, B, and C, hypothesis 1 would be the preferred one. But using stratigraphy to choose between hypotheses 1 and 2 is, in our opinion, highly problematic, because all it would take to completely overturn hypothesis 1 would be the discovery of a single fossil lower in the stratigraphic sequence (i.e., taxon C below taxon A). Phylogenetic trees tend to be much more stable, because the discovery of a single incongruent piece of information rarely, if ever, overturns a well-supported cladogram. Thus a relatively high degree of completeness of the fossil record (or at least a constant rate of incompleteness in all clades) would be necessary for stratigraphy to be a reliable test of phylogenetic hypotheses, especially for groups that may have diversified rapidly over a relatively short period of time.

In our opinion, we cannot rely on the fossil record for absolute age range estimates of taxa. This is an old warning, but one that is nevertheless worth restating. The nonoccurrence of a taxon in certain strata of a specific area does not mean that the taxon did not exist elsewhere or at other times. Apparent nonoccurrence of a taxon in known fossil deposits can also mean one of the following: (1) individuals of the taxon decomposed or were otherwise destroyed or made unrecognizable before fossilization (taphonomy, discussed by Fisher); (2) the taxon was not preserved because it occupied a different geographic or ecological realm whose sediments were not preserved; (3) the sediments containing the fossil were eroded or otherwise destroyed before discovery by humans; (4) the taxon was preserved but has not yet been discovered. All of these reasons, and perhaps others, result in gaps in the fossil record. Although we have no way of identifying the exact size or consistency of gaps in the fossil record if they represent a pre-Holocene endpoint of the stratigraphic range, we can identify the size of gaps when they occur somewhere between the first and last known occurrence of a monophyletic group. Intuitively, one might expect that these gaps would decrease with continued research and new discoveries. But so far, the old problem of gaps has become more and more extensive and clearly defined with increased phylogenetic resolution and discovery of certain new taxa. Fishes and marine tetrapods, for example, should have a relatively complete fossil record compared to that of land-dwelling vertebrates, because of their aquatic mode of life and proximity to the sediments that became sedimentary rocks. Yet well-known gaps in the fossil record of fishes are remarkable. For example, there are no known coelacanth fishes between Late Cretaceous time and the present. In fact, prior to the discovery of the one living coelacanth species in 1938, the group was thought to be extinct since the Cretaceous. The discovery of the living species added a previously unknown 70-million-year gap to the fossil record of coelacanths. Similarly, there are currently no known lampreys between mid-Pennsylvanian time and the present (about a 300-million-year gap), and prior to 1968 there were *no* known fossil lampreys. (They are now known from Pennsylvanian and Mississippian time [Bardack and Zangerl, 1968; Janvier and Lund, 1983]). Nearly all extant families of fishes and tetrapods that have been analyzed cladistically and that have extensive fossil records (e.g., with records going back at least as far as Cretaceous time) show significant gaps of tens of millions of years or more in their stratigraphic distribution. Consequently, there must be a significant margin of error when calculating an absolute age range for a taxon based on stratigraphy. This margin of error seems too high to allow stratigraphic data an important role in phylogeny reconstruction.

Fisher's analysis of biogeographic patterns (Chapter 6, this volume) likewise involves "assessing where taxa are *not* located in space-time, as well as where they are." The absence of taxa in space is as much negative evidence as is their absence in a given stratigraphic layer. If relationships of taxa (e.g., see Grande, Chapter 4, this volume) rather than geographic range data are used to determine area relationships, we can move beyond the use of "presence/absence" data and decouple studies of historical biogeography from negative evidence. For example, and referring to Figure 2 in Grande, family ABC indicates a close area relationship between Areas 1 and 2 not because it is absent from Area 3, but because of the relationships of its subtaxa. The family is present in all three areas and forms a pattern of relationship independently repeated by other families of organisms. If most taxa within a monophyletic group independently show the same biogeographic pattern, it can minimize the probability that the biogeographic similarity between Areas 1 and 2 is merely the result of lack of preservation or massive extinction of taxa in Area 3.

The use of congruence alone to identify putative homology leaves some morphologists and developmental biologists dissatisfied. Wake opens Chapter 7 by drawing a distinction between morphologists (including functional morphologists) and systematists, believing that morphologists may pursue a research program independent from systematic considerations. If, however, morphologists address the problem of evolutionary transformation of a complex morphological structure, she argues they must do so within a rigorous phylogenetic framework. Wake discusses the use that functional morphologists make of patterns identified by other authors and on the basis of other characters (i.e., mapping functional morphological complexes on pre-existing cladograms). She further argues that evolution of functional morphological complexes must be studied with respect to a well-corroborated pattern, with cladograms used as important constraining factors in the inference of process explanations. The same methodological rigor has recently been advocated for the study of evolution of ecosystems and/or behavior (Brooks and McLennan, 1991).

Structural complexity is an inherently tricky concept (Webster, 1984; Bonner, 1988), as are natural selection and selection pressure. To hypothesize evolutionary transformation of functional morphological complexes in terms of "plausibility" with respect to "selection pressures", "biological roles", and "adaptive values" (Bock and von Wahlert, 1965) is a futile exercise, in our opinion. It is our view that any hypothesis of transformation remains empirically empty if not constrained by a rigorous phylogenetic framework. It is imperative that functional morphologists map functional complexes on a pre-existing cladogram to study their transformation.

Intuitively obvious as this research program of functional morphology would appear, it nevertheless carries with it the threat of circularity, as emphasized by Wake. It is one thing to map characters on a pre-existing cladogram to study their evolution, but it is another thing to then allow the same characters to contribute to the process of cladogram construction. By mapping a functional morphological complex on a pre-existing cladogram, the steps of transformation of this complex are put into some sequential order, but the structural complex itself does not contribute to the test of the cladogram. It is assumed *a priori* that the cladogram reflects the best estimate of phylogenetic relationships available, and that the functional complex evolved within that phylogenetic framework. We might ask, however, what might happen if the structural complex under consideration were subjected to the test of congruence along with all the other characters on which the pre-existing cladogram is based. Such would be required by the principle of total evidence (discussed above), and congruence might indicate that with the addition of the functional complex to the data, the topology of the cladogram would change. In this case previous notions about the evolutionary history of the functional morphological complex would be falsified. Wake addresses this problem when she urges that besides mapping structural complexes on the cladogram, the functional morphologist should also investigate the systematic potential of such structural complexes for phylogenetic analysis.

Using functional morphological complexes to study evolutionary transformations entails two further problems also dealt with by Shubin (Chapter 8, this volume), namely character interdependence and homology. As the term suggests, a functional morphological complex is composed of several structural components integrated into a common functional context. As the utility of such a structural complex for cladogram building is investigated, the question arises as to whether the whole functional complex should each be coded as a single character, or whether its component elements should be coded as separate characters (Barel, 1984). Wake raises the issue of character interdependence, and looks for solutions to that problem, e.g., in developmental biology. As discussed by Shubin, causal developmental sequences (Alberch, 1985) may result in structural complexes whose components will always occur together, owing to developmental integration. "Developmental constraints" would thus seem to provide an empirical clue to character interdependence prior to phylogenetic analysis. But we ask, how is it possible to discover "developmental constraints" or to falsify conjectures of such constraints? The first step would be to look for perfect congruence among some set of characters. Such perfect congruence would indicate structural invariance calling for a causal explanation. "Developmental constraints" have not yet materialized as an empirical research program of developmental biology. Rather, "developmental constraints" remain a descriptive concept for structural invariance, that is, perfect character congruence. And if phylogenetic analysis, that is, the test of congruence involving a suite of additional characters, were to document incongruence among constituent elements of a structural complex, the obvious answer would be that developmental mechanisms had changed, that evolution resulted in a change of developmental constraints. In our view, it is not developmental biology that can inform us about character interdependence or independence. Instead, it is pattern analysis that informs us about the evolution of developmental integration. It is homology as it emerges from the test of congruence that teaches us about the evolution of development. To investigate evolutionary transformations of a developmental or functional morphological complex by mapping it on a pre-existing cladogram makes the *a priori* assumption that the developmental or functional morphological complex under consideration is homologous throughout the group being studied. How can we know if it is or is not?

In the context of her distinction between morphologists and systematists, Wake (Chapter 7, this volume) proposes that there are means to postulate homology for morphological structures (such as the tentacle in caecilians) independent of the test of congruence. Similarly, Shubin (Chapter 8, this volume) discusses the role of developmental biology in the initial conjecture of relations of homology. Wake and Shubin both distinguish the notion of "biological homology" from the less inclusive concepts of "phylogenetic homology" and "iterative [serial] homology". Biological homology, first introduced by Wagner (1989), "refers to biological mechanisms rather than to genealogical connections alone" (p. 55), and thus implicitly includes phylogenetic homology together with iterative (serial) homology and perhaps other non-phylogenetic components. The concept of biological homology proposes that common ancestry would not be necessary for two structures to be homologous. To us, the addition of non-phylogenetic components to the concept of homology renders the concept of homology irrelevant to comparative biology and phylogenetic studies. Suppose two putatively homoplastic structures appear to be exactly alike, and originate through similar ontogenetic transformations: those structures could not be told apart (i.e., identified as homoplasies) by any criterion except congruence. With respect to discernible biological properties, these structures would be identical, and hence would qualify as "biological homologies". Wagner (1989, p. 62), cited by both Wake and Shubin (this volume), defined biological homologies in the following way: "Structures from two individuals or from the same individual are homologous if they share a set of developmental constraints, caused by locally acting self-regulatory mechanisms of organ differentiation. These structures are thus developmentally individualized parts of the phenotype". But we ask, what does it mean that structures and developmental constraints are shared? What is shared similarity in two unrelated groups of organisms but homoplasy of structure as well as homoplasy of developmental pathways or constraints, and thus *analogy* rather than homology? The test for homology is congruence with respect to other characters, not commonality of developmental pathways or developmental constraints. Indeed, Shubin emphasizes that development may be used in the initial conjecture of homology ("biological homology"), but that these conjectures of homology still need to be tested by congruence. But biological homology becomes testable by congruence only when the non-phylogenetic components are removed from it, and when they are removed, the resulting homology concept becomes indistinguishable from the phylogenetic homology. Thus, to us the "biological homology" concept is empirically empty, and of little or no practical use to evolutionary studies.

The attempt to identify homology by means other than the test of congruence, by shared developmental constraints or pathways, leads to the structuralists' research program commented upon by both Wake and Shubin (Chapters 7 and 8, this volume). Structuralists pursue an ahistorical approach to comparative biology. They contend that the Darwinian theory of evolution does not explain apparent domains of structural invariance. If evolution is, as Darwin proposed, a matter of variation and natural selection, and if mechanisms of variation and natural selection are necessary and sufficient to explain the diversification of taxa over time, then one would expect fundamental variability of organisms in all aspects of their organization. The fact that organisms can be classified in subgroups within groups, according to the structuralist, indicates domains of invariance in the structure of organisms. Invariance, as opposed to random variability, remains unexplained under the Darwinian paradigm. Structuralists maintain that structural invariance must have its cause in a realm outside Darwinian theory, and morphological invariance in development must be governed by time-independent laws of structure, comparable to the law of gravity in physics. If we assumed that such time-independent laws of structure exist, it would seem possible to classify organisms not on the basis of shared history, i.e., shared ancestry, but rather on the basis of shared structural laws. The result would be some kind of periodic table of organisms, a rational classification of organisms, based not on history, but on logical relations of form. However, it has long been recognized that morphology is only partially the outcome of genetic information and structural laws, because it is also heavily influenced by epigenetic (i.e., environmental) factors. Ho (1988), as discussed by Wake (Chapter 7, this volume), deduces that because morphology is vulnerable to environmental and other influences, it is a poor guide to genealogy. Accordingly, Ho contends that genealogy (i.e., phylogeny) should be based on DNA data alone, since those data presumably are not subject to epigenetic modification. She feels that morphology, on the other hand, should be viewed as an outcome of structural laws of form, epigenetically modified by environmental factors. She also indicates that she believes that the study of morphology should proceed with reference to a phylogenetic framework provided by DNA alone.

Implicit in the structuralist research program is a fundamental decoupling of molecular and morphological evolution, and the idea that morphology cannot be used in the study of genealogy because it is the outcome of time-independent laws of structure and modifying environmental influences. This is reflected in Wake's quotation from Ho (1988), who asserted that a "basic assumption" of phylogenetic systematics is that morphological data are congruent with phylogeny. We believe that no such assumption is necessary to justify the use of morphology in pattern reconstruction. Indeed, to claim that morphological data are better or worse than any other form of data to determine phylogeny implies independent knowledge of "the true phylogeny". To our knowledge, no such claim has ever been voiced by proponents of cladistic analysis. The search is for congruence of character distributions, whether morphological or molecular. There is controversy about methods of character analysis (taxonomic congruence versus total evidence methods of analysis), but there is agreement that common descent is a hypothetical explanation of the hierarchical patterns discovered by congruence. Phylogeny is not "known truth" that can be used to compare patterns supported by morphological data with those supported by molecular data for degree of accuracy.

While there can be no serious doubt that laws of structure exist, and that morphology is also influenced by environmental factors, the relevant question in evolutionary biology is, what do we make of the order we perceive in nature? Following the structuralist research program, incongruence between a pattern based on morphology and one based on molecular data would be expected, and morphological data would not be used in the study of genealogy. But we find that the structuralist research program fails to provide an explanation for the hierarchical order of morphological data that we observe in nature. As long as the study of evolutionary processes must start with the search for systematic pattern, and as long as the search for such patterns is successful, there is no room for various notions of shared similarity outside the constraints of the taxic hierarchy. It is order, not disorder, which is in need of a process explanation.

So we end this volume by identifying what we see as a key theme uniting many different perspectives expressed in this volume: the search for order in nature on the basis of congruence. Science and philosophy are both heavily woven into studies of the apparent orderliness of nature (systematics), and historical explanations for that apparent order (e.g., evolutionary theory, historical biogeography, connections between ontogeny and phylogeny). Systematic patterns and theories of evolutionary process continue to form the most basic reference point for all biological sciences. However, we see the connections between process theory and systematic patterns in need of much further development. Within the various subdisciplines of biology, there are still many different perspectives on how to study evolution, pattern, and process; but for us, the basic element that seems to be most important is the support of a hierarchy of subordinated three-taxon statements by congruent data. Thus, we will probably be labeled "pattern cladists". Data congruence is what allows us to investigate homology, phylogeny, and biogeographic history in the most consistent (and therefore the most objective) manner. In order to search for elements of consistency in nature, it is necessary to find methods of analysis that are also consistent. This is not to imply that nature has always acted in a consistent manner; but elements of consistency are what we, as scientists, are best equipped to discover empirically. And without an empirical component to our endeavors, it is not science that we are practicing.

Acknowledgments

We thank Gary Nelson, Colin Patterson, and Mario de Pinna for reading this chapter and providing us with a number of helpful comments that improved the content.

References

Alberch, P. 1985. Problems with the interpretation of developmental sequences. *Systematic Zoology*, 34: 46-58.

Bardack, D., and R. Zangerl. 1968. First fossil lamprey: a record from the Pennsylvanian of Illinois. *Science*, 1962: 1265-1267.

Barel, C. D. N. 1984. Form-relations in the context of structural morphology: the eye and suspensorium of lacustrine Cichlidae (Pisces, Teleostei). *Netherlands Journal of Zoology*, 34: 439-502.

Beatty. J. 1982. Classes and cladists. *Systematic Zoology*, 31: 24-34.

Bock, W. J., and G. von Wahlert. 1965. Adaptation and the form-function complex. *Evolution*, 19: 269-299.

Bonner, J. T. 1988. *The Evolution of Complexity.* Princeton, NJ: Princeton University Press. 260 pp.

Brady, R. H. 1982. Dogma and doubt. *Biological Journal of the Linnean Society*, 17: 79-96.

Brady, R. H. 1983. Parsimony, hierarchy, and biological implications. In N. I. Platnick and V. A. Funk (Eds.), *Advances in Cladistics*, Vol. 2. *Proceedings of the Second Meeting of the Willi Hennig Society*, 49-60. New York: Columbia University Press.

Brady, R. H. 1985. On the independence of systematics. *Cladistics*, 1: 113-126.

Brooks, D. R. 1990. Parsimony analysis in historical biogeography and coevolution: methodological and theoretical update. *Systematic Zoology*, 39: 14-30.

Brooks, D. R., and D. A. McLennan. 1991. *Phylogeny, Ecology, and Behavior.* Chicago: The University of Chicago Press. 434 pp.

Cracraft, J. 1988. Deep-history biogeography: retrieving the historical pattern of evolving continental biotas. *Systematic Zoology*, 37: 221-236.

Darwin, C. 1859. *On the Origin of Species by Means of Natural Selection, or the Preservation of Favoured Races in the Struggle for Life.* London: John Murray. 502 pp.

de Queiroz, K. 1988. Systematics and the Darwinian Revolution. *Philosophy of Science*, 55: 238-259.

Dobzhansky, Th., 1970. *Genetics of the Evolutionary Process.* New York: Columbia University Press. 505 pp.

Dobzhansky, Th., F. J. Ayala, G. L. Stebbins and J. W. Valentine. 1977. *Evolution.* San Francisco: W. H. Freeman and Co. 572 pp.

Eldredge, N. 1971. The allopatric model and phylogeny in Paleozoic invertebrates. *Evolution*, 25: 156-167.

Eldredge, N., and J. Cracraft. 1980. *Phylogenetic Patterns and the Evolutionary Process.* New York: Columbia University Press. 349

pp.
Eldredge, N., and S. J. Gould. 1972. Punctuated equilibria: an alternative to phyletic gradualism. In T. J. M. Schopf (Ed.), *Models in Paleobiology*, 82-115. San Francisco: W. H. Freeman and Co. 250 pp.
Evans, S. E., and M. K. Hecht. 1993. A history of an extinct reptilian clade, the Choristodera: longevity, Lazarus-taxa, and the fossil record. *Evolutionary Biology*, 27: 323-338.
Fisher, D. C. 1992. Stratigraphic parsimony. In W. P. Maddison and D. R. Maddison, *MacClade, Version 3*, 124-129. Sunderland, MA: Sinauer Associates Inc. Publishers. 398 pp.
Gilmour, J. S. L. 1961. Taxonomy. In A. M. MacLeod and L. S. Cobley (Eds.), *Contemporary Botanical Thought*, 27-45. Chicago: Quadrangle Books.
Goldschmidt, R. 1940. *The Material Basis of Evolution*. New Haven: Yale University Press.
Gould, S. J., and N. Eldredge. 1977. Punctuated equilibria: the tempo and mode of evolution reconsidered. *Paleobiology*, 3: 115-151.
Grande, L. 1985. The use of paleontology in systematics and biogeography, and a time control refinement for historical biogeography. *Paleobiology*, 11(2): 1-11.
Hennig, W. 1966. *Phylogenetic Systematics*. Urbana: University of Illinois Press. 263 pp.
Ho, M.-W. 1988. How rational can rational morphology be? A post-Darwinian rational taxonomy based on a structuralism of process. *Revista di Biologia*, 81: 11-55.
Humphries, C. J., and L. Parenti. 1986. Cladistic biogeography. *Oxford Monographs in Biogeography*, 2: 1-98.
Janvier, P., and R. Lund. 1983. *Hardistiella montanensis* n. gen. et sp. (Petromyzontida) from the Lower Carboniferous of Montana, with remarks on the affinities of the lampreys. *Journal of Vertebrate Paleontology*, 2(4): 407-413.
Kluge, A. G. 1985. Ontogeny and phylogenetic systematics. *Cladistics*, 1(1): 13-27.
Kluge, A. G. 1989. A concern for evidence and a phylogenetic hypothesis of relationships among Epicrates (Boidae, Serpentes).

Systematic Zoology, 38: 315-328.
Mayr, E. 1942. *Systematics and the Origin of Species*. Reprint 1964, New York: Dover Editions. 334 pp.
Mayr, E. 1963. *Animal Species and Evolution*. Cambridge, MA: Belknap Press at Harvard University Press. 797 pp.
Mayr, E. 1982. *The Growth of Biological Thought*. Cambridge, MA: Belknap Press at Harvard University Press. 974 pp.
Nelson, G. 1989. Cladistics and evolutionary models. *Cladistics*, 5: 275-289.
Nelson, G., and C. Patterson. 1993. Cladistics, sociology and success: a comment on Donoghue's critique of David Hull. *Biology and Philosophy*, 8: 441-443.
Nelson, G., and N. Platnick. 1981. *Systematics and Biogeography, Cladistics and Vicariance*. New York: Columbia University Press. 567 pp.
Norell, M. A., and M. J. Novacek. 1992a. The fossil record and evolution. Comparing cladistic and paleontological evidence for vertebrate history. *Science*, 255: 1690-1693
Norell, M. A., and M. J. Novacek. 1992b. Congruence between superpositional and phylogenetic pattern: comparing cladistic patterns with fossil records. *Cladistics*, 8: 319-337.
Ospovat, D. 1981. *The Development of Darwin's Theory. Natural History, Natural Theology, and Natural Selection, 1838-1859*. Cambridge: Cambridge University Press. 301 pp.
Otte, D., and J. A. Endler (Eds.). 1989. *Speciation and its Consequences*. Sunderland, MA: Sinauer Associates Inc. 679 pp.
Patterson, C. 1980. Cladistics. *Biologist*, 27: 234-240.
Patterson, C. 1982. Morphological characters and homology. In K. A. Joysey and A. E. Friday (Eds.), *Problems of Phylogenetic Reconstruction*, 21-74. London: Academic Press.
Platnick, N. I. 1977. Cladograms, phylogenetic trees, and hypothesis testing. *Systematic Zoology*, 26: 438-442.
Popper, K. R. 1963. *Conjectures and Refutations. The Growth of Scientific Knowledge*. New York: Harper Torchbooks. 417 pp.
Ridley, M. 1986. *Evolution and Classification. The Reformation of*

Cladism. London: Longman., 201 pp.

Rieppel, O. 1988. *Fundamentals of Comparative Biology*. Basel: Birkhäuser Verlag. 202 pp.

Rieppel, O. 1992. Homology and logical fallacy. *Journal of Evolutionary Biology*, 5: 701-715.

Rieppel, O. 1994. The role of paleontological data in testing homology by congruence. *Acta Palaeontologica Polonica*, 38(3): 295-302.

Rosenberg, A. 1985. *The Structure of Biological Science*. Cambridge: Cambridge University Press. 281 pp.

Simpson, G. G. 1944. *Tempo and Mode of Evolution*. New York: Hafner Publishing Co. 237 pp.

Sober, E. 1988. The conceptual relationship of cladistic phylogenetics and vicariance biogeography. *Systematic Zoology*, 37: 245-253.

Wagner, G. P. 1989. The biological homology concept. *Annual Review of Ecology and Systematics*, 20: 51-69.

Webster, G. 1984. The relations of natural forms. In M.-W. Ho and P. T. Saunders (Eds.), *Beyond Neo-Darwinism*, 193-217. London: Academic Press. 376 pp.

Wiley, E. O. 1981. *Phylogenetics. The Theory and Practice of Phylogenetic Systematics*. Chichester: John Wiley and Sons. 439 pp.

Wiley, E. O. 1988a. Vicariance biogeography. *Annual Review of Ecology and Systematics*, 19: 513-542.

Wiley, E. O. 1988b. Parsimony analysis and vicariance biogeography. *Systematic Zoology*, 37: 271-290.

Wiley, E. O., D. Siegel-Causey, D. R. Brooks and V. A. Funk. 1991. *The Compleat Cladist. A Primer of Phylogenetic Procedures*. Special Publication No. 19. Lawrence: University of Kansas Museum of Natural History. 158 pp.

Glossary

Lance Grande and Olivier Rieppel

Department of Geology
Field Museum of Natural History
Chicago, Illinois, 60605

WE ADDED THIS GLOSSARY to help the general reader digest some of the more specialized sections of this volume. We trust that the advanced systematist or evolutionary biologist will not find the simplified treatment of certain terms ("cladistics", "species", "historical biogeography" or "Darwinism", for example) annoyingly brief, but it was not our intent to provide a comprehensive historical account of all terms. Because of the complex history of some of these terms today, there is no way to do them justice in a glossary. We also provided suggested references for a number of these terms.

The terms in this glossary are briefly explained with regard to their relevance to systematics and evolutionary studies, although they may also have other meanings and different contexts. The locations where many of these terms are used in this volume are given in the subject index.

Algorithm. An explicit method of solving a mathematical problem. In systematic studies, algorithms usually consist of mathematical methods to find a regular pattern of data, and are used in computer programs to analyze comparative data.

Anagenesis. Evolutionary change that does not involve branching (as opposed to cladogenesis). Usually used to refer to gradual changes

Copyright © 1994 by Academic Press, Inc.
All rights of reproduction in any form reserved.

within a species or a lineage. See Speciation, Cladogenesis.

Analogy. A similarity not due to common ancestry; homoplastic similarity. Examples: the tail of a teleost fish could be considered to be analogous with the tail of a whale, the eye of a squid could be considered analogous with the eye of a human. See Homoplasy, Homology.

Ancestral character state. A character condition thought to be primitive for a monophyletic group based on outgroup comparison. For example, the taxonomic groups most closely related to snakes have legs; consequently we would assume that snakes descended from an organism that had legs. See Outgroup comparison method.

Ancestral taxon. A term used by some authors (e.g, Wiley et al., 1991) to refer to a species that gave rise to one or more descendant species.

Apomorphy or **Apomorphic character.** A character hypothesized to be uniquely derived for (i.e., diagnostic of) a particular monophyletic taxon. In evolutionary terms, an apomorphy would be a peculiar feature shared by the members of a monophyletic taxon that was inherited by each of those members from a hypothetical common ancestor. Presumably, this feature originally evolved in the common ancestor. Some authors divide the term apomorphy into two subcategories: "autapomorphy" for terminal taxa (distalmost branches of a given cladogram), and "synapomorphy" for monophyletic groups of taxa (branch points of a given cladogram). Apomorphy is equivalent to phylogenetic homology or derived character. Examples: the presence of feathers for Aves (birds), the presence of jaws for Gnathostomata (vertebrates with jaws).

A posteriori. Reasoning based on observation or empirical data; reasoning after the fact.

A priori. Reasoning based on theory or assumption rather than observation or empirical evidence; reasoning before the fact.

Archetype. A basic body plan modeled after a number of general morphological characters common to a group of taxa; used by Darwin and other early evolutionists to construct a model of a hypothetical ancestor for a given taxon.

Area relationship. Used in historical biogeography, this term refers to the relationships between geographic areas based on the interrelationships of the taxa inhabiting them.

Autapomorphy. See Apomorphy.

Binary character. See Character.

Biogenetic law. See Haeckel's Law.

Biogeography. The study of the relationship between geography and biology. Some of the general ideas and principles of biogeography date back at least as far as the early part of the 19th century (Nelson, 1978b). The form of biogeography most relevant to the study of evolution and systematics is historical biogeography. See Historical biogeography. For further reading, see Nelson and Rosen (1981), Brown and Gibson (1983), and Humphries and Parenti (1986).

Biological homology. A term that redefines homology "on a mechanistic rather than a genealogical basis" (Wagner, 1989, p. 54). Includes both phylogenetic homology and iterative homology. Discussed further in Wagner (1989) and Roth (1984). See Phylogenetic homology, Iterative homology.

Biostratigraphy. The paleontological aspects of stratified (bedded or layered) rocks (usually sedimentary rocks). For example, the use of fossil species or assemblages of fossil taxa to date a layer of rock, or the use of the vertical order of occurrence of fossil taxa to influence the reconstruction of phylogeny (i.e., considering the oldest fossils to be the most primitive, morphologically).

Biota. The entire community of organisms found in an area.

Bootstrapping. A numerical technique of resampling (Efron, 1979, 1982) adapted by Felsenstein (1985) to calculate indices of support for branching points (nodes) of cladograms or phenograms. Simulated data sets are generated by randomly sampling the original character set. Each simulated data set is made to have the same number of characters as the original data set. Consequently, in most cases some characters from the original data set are used more than once and some not at all. Many (100 or more) of these simulated data sets are each analyzed cladistically or phenetically. An index of support for each node of the original cladogram or phenogram is calculated based on the percentage of simulated data sets that show it. For additional reading, see Felsenstein (1985), Lanyon (1987), Hillis and Moritz (1990), Sanderson (1989), and Dodds (1986). For a recent criticism of this technique, see Kluge and Wolf (1993). Although bootstrapping techniques can also be used to resample taxa (rather than characters), jackknifing is the method most often used for resampling taxa. See Jackknifing.

Branch-and-bound method. An algorithm first described by Hendy and Penny (1982) that helps to reduce the number of phylogenetic trees that must be calculated (and thus reduce computer time required) in a search for the most parsimonious pattern or tree. See Kitching (1992, pp. 65-66) and Swofford and Olsen (1990) for further reading on this method. This method does not produce all possible trees, but it does find the most parsimonious tree or trees, and it streamlines a search when exhaustive search is not practical. See Exhaustive search method.

Branching diagram. A diagram of taxa connected by lines (such as a cladogram, phenogram, network or tree) that expresses analytical results and/or a theory of relationships between taxa. See also Tree.

Character. Any feature that is an observable part or attribute of an organism. Characters that diagnose groups of organisms are conjectures of homology. *Binary characters*, or two-state characters, are those with only one derived state (e.g., the presence of feathers versus their absence). *Multistate characters* have more than one derived state (e.g., the presence of 3 or 4 toes in the ingroup and the presence of 5 toes in the outgroup). Some systematists prefer to use binary characters in their analyses, and a multistate character can be converted into a number of binary characters. For further reading on this subject see Wiley et al. (1991) and Forey et al. (1992).

Character distribution pattern. See Cladogram.

Character interdependence. Interdependent characters are those which always appear together (i.e., they are thought to be functionally or developmentally linked). Consequently, two such characters might not be treated as different characters in an analysis. For example, simple variation of eye size can influence a plethora of cranial characters (skull bones surrounding the eye), all of which could be scored separately, but all of which may in fact reflect only a single character, i.e., eye size. The question of whether two particular characters are independent or interdependent is sometimes controversial. See Wake (Chapter 7).

Character polarity. See Polarity.

Clade. A monophyletic group (as monophyletic is used here). See Monophyletic group.

Cladistics. Currently the most widely used method of investigating evolutionary relationships among organisms. It can also be used to investigate the relationships between biogeographic areas (see Chapter 4). Cladistic methodology looks at the relative relationships of taxa based on analysis of empirical data (e.g., see Fig. 1, p. 64) and does not generally attempt to identify specific ancestors. Cladistics attempts to use synapomorphic characters to reconstruct genealogical relationships among taxa, and it assembles taxa based on relative relationships of monophyletic groups. Unlike phenetics, which groups taxa by overall similarity, cladistics attempts to use only putatively "positive" character information. For example, the group Invertebrata is not a cladistic group, because it is based on negative character information (the absence of vertebrae). Secondary losses (characters putatively lost phylogenetically) are not treated as negative characters because losing something is not the same as never having had something. For example, sturgeons have no vertebrae, but scientists believe, based on an abundance of other characters, that sturgeons have an ancestor that did have vertebrae. It is impossible to thoroughly discuss a term like this in a glossary, so we encourage the general reader interested in more information to go to a number of excellent sources such as Wiley et al. (1991), Eldredge and Cracraft (1980), Wiley (1981), Schoch (1986), Forey et al. (1992), and Nelson and Platnick (1981). Also see Apomorphy, Cladogram, Synapomorphy.

Cladistic pattern. See Cladogram.

Cladogenesis. A process theory of lineage multiplication involving branching (splitting) and divergence between species over time. See Anagenesis, Speciation.

Cladogram. A type of branching diagram that uses a hierarchical organization of data to construct a putative phylogeny of taxa (for example, see Chapter 4, Fig. 1). The characters on a cladogram that exhibit congruence are thought to be indicative of evolutionary relationship, and independently derived for monophyletic taxa. These characters are termed apomorphies or synapomorphies. For example, feathers are thought to be uniquely derived for birds (and their presence in all birds is hypothesized to be due to inheritance from a common ancestor). See Apomorphy, Cladistics, Hierarchy, Parsimony.

Common ancestor. The hypothetical ancestor of a monophyletic group of organisms. On a cladogram, the hypothetical common ancestors are at the branch nodes. See also Tree.

Congruence. Congruent data are those data that fit together with no conflict. The degree of congruence is dependent on the percentage of the total number of characters in a data set that are congruent with each other. See Chapter 1.

Consensus tree. A tree that is the summary of the branching information from two or more equally parsimonious (or nearly equally parsimonious) trees or from independent data sets. Consensus techniques were originally developed to accommodate multiple data sets or multiple trees that are each based on a different data set (Adams, 1972). Consensus techniques were eventually used also to address the problem of multiple equally parsimonious (or nearly parsimonious) trees for a single data set (Barrett et al., 1991). The use of consensus methods in phylogenetic analysis is controversial (e.g., see Barrett et al., 1991; Nelson, 1993). See Parsimony.

Consistency index. A measure of congruence for a set of characters on a branching diagram between 0 and 1, with 1 meaning there is no homoplasy (i.e., the data fit the branching diagram with no conflict). The consistency index for a binary character is a ratio of 1 over the number of times the character appears on a cladogram, so if the character appears only once, the CI is 1/1, or 1. If the same character is homoplastic and shows up independently on two different branches, the CI is 1/2, or .5.

Convergence. See Homoplasy.

Darwinian theory. See Darwinism, Neodarwinism.

Darwinism. A school of evolutionary thought that stipulates a strict adherence to Darwinian mechanisms of change in evolutionary explanation. Darwinian mechanisms are those of variation and natural selection working gradually at the level of populations. Thought by a number of authors to be a nontestable concept, and hence not scientific by some definitions (Pauper, 1976; Patterson, 1978; Brady, 1980, 1982). See Population, Gradualism, Natural selection, Neodarwinism.

Data matrix. A table of data (e.g., see Chapter 5, Table 2). Usually, in a data matrix, the columns indicate character states and the rows represent taxa.

Derived character. See Apomorphy.

Developmental biology. The study of change in an organism from conception to death.

Developmental constraints. A theoretical concept specifying that certain morphological conditions are unable to develop owing to causal limitations on the possible ways an organism can develop throughout its life cycle. For example, there are no known cases of a hand in which there are more phalanges in digits I or V than in digits II through IV, and this is thought by some to be the result of form-generating mechanisms of embryonic development that prevent certain structures from differentiating. Certain structures such as the phalanges discussed above are thought to be restricted to "domains of invariance". For further reading, see Holder (1983).

Developmental pathway. The sequence of modification that a specific trait or structure undergoes during development until the mature stage is reached.

Diagnose (with regard to taxa). To define a group so it can be distinguished from all other groups within a taxon.

Dispersal. With regard to historical biogeography, the movement of a taxon outside its previous geographic range through migration of individuals.

Distance analysis. A method of analysis that uses degree of dissimilarity between genes to produce a branching diagram of taxa. Various distance methods are reviewed in Nei (1991). See Branching diagram.

Dollo's Law. The notion that a complex morphological structure cannot re-evolve once it has been lost through evolution. The modern use of the term is extensively discussed by Gould (1970). In cladistic analysis, Dollo parsimony weights acquisition of evolutionary novelties more highly than their secondary loss.

Domains of structural invariance. See Developmental constraints.

Epistemology. A branch of philosophy that studies the origin, nature and boundaries of human knowledge.

Empirical. Based on experiment and/or descriptive data.

Evolution. Unidirectional (non-cyclic) change.

Evolutionary homology. See Phylogenetic homology.

Evolutionary process. A historical causal explanation for the apparent orderliness and systematic patterns of the biological world.

Exhaustive search method. An algorithm outlined by Swofford and Olsen (1990) that evaluates all possible strictly bifurcating trees for a given data set to find the most parsimonious tree or trees. This method rapidly becomes cumbersome with the addition of taxa (e.g., for 20 taxa there are 220,000,000,000,000,000,000 possible trees) and is not practical for data sets of over 11 taxa. Thus, methods that streamline the search for parsimony are also available (e.g., see Branch and bound method). This method is described in detail in Kitching

(1992, pp. 63-65). See Algorithm, Parsimony.

Functional morphological complexes. A group of morphological characters that appear to be linked or interdependent on each other to serve a complex function such as flight, swimming, etc.

Gaps (in the fossil record). Geological time segments where particular taxa are missing. For example, at the time this volume was written there were fossil lampreys known from Pennsylvanian time (about 290 million years before present) and lampreys alive today, but no fossil lampreys between Pennsylvanian time and modern time.

Gaps (in gene sequences). Presumed insertion and deletion events. These gaps are represented by editing symbols inserted into gene sequence data in order to help align homologous (orthologous) segments for comparison.

Gaps (in morphology). Missing segments in a theoretical morphological continuum; a discontinuity in morphological transformation indicated by phylogeny.

Genealogy. The recorded history of descent from an ancestor or ancestors; lineage; ancestor-descendant relationship.

Genotype. The set of genes possessed by an organism that interact with environment to produce the phenotype (or attributes) of an organism. Also used by some authors to refer to the genetic information of a single, or a few, specific genes. See Phenotype.

Grade. An artificial (i.e., non-monophyletic) taxon. The term is usually used to indicate a hypothetical level of "evolutionary progress" (e.g., Reptilia or Pisces in their traditional usage). This term is more relevant to paleoecological studies than to evolutionary studies.

Gradual evolution. A theory that describes evolutionary process as a gradual morphological transition, without sudden change. This is opposed to theories of discontinuous change, such as macromutations, quantum evolution, or punctuated equilibrium. See Macromutations, Quantum evolution, Punctuated equilibrium.

Gradualism. See Gradual evolution.

Haeckel's Law. The notion that "ontogeny recapitulates phylogeny" (=the biogenetic law). This was not really a law, but a generalization based on an observed trend. This concept was argued against by de Beer (1958) and others, and redefined in a revised form by Nelson (1978a). See Recapitulation.

Hardy-Weinberg Principle. A statistical model developed by population geneticists for interpreting the genetic makeup of a population at equilibrium and under certain conditions. This equation is also referred to as the Hardy-Weinberg Law, the Hardy-Weinberg theorem, or the Hardy-Weinberg equilibrium principle by various authors (e.g., Futuyma, 1979; Murphy et al., 1990). This equation states that in a randomly mating population of infinite size where there is no immigration, emigration, selection, or mutation, a stable equilibrium will be reached where the genotype frequencies will be $AA=p^2$, $Aa=2pq$, and $aa=q^2$, where p is the frequency of allele A and q is the frequency of allele a.

HENNIG 86 (version 1.5). A computer program developed by Farris (1988) to calculate the most parsimonious pattern or tree from a data set.

Hierarchy. An organization of things arranged in order of rank; a pattern of subsets nested within larger sets.

Historical biogeography. The study of the relationships between organisms and geographic areas through time. Two major subcategories of historical biogeography are ecological biogeography and phylogenetic/vicariance biogeography. Ecological biogeography is the study of "the distribution and history of populations, species, and biotic communities as members of ecosystems" (Brundin, 1981). Phylogenetic/vicariance biogeography uses a combination of the cladistic relationships of monophyletic taxa and the geographic ranges of those taxa to investigate geographic area relationships (e.g., see Chapter 4, Fig. 2). Traditionally, most biogeographers used only range data for biogeographic studies, but the addition of a cladistic analytical component to the method can decrease the probability that a biogeographic pattern is only an artifact of extinction and the incompleteness of the fossil record. See Chapters 4 and 9. See Area relationships, Biogeography, Range data.

Homology. The use of this term today is variable, and thus confusing, but it generally refers to some aspect of "sameness" (e.g., structures that match each other in origin, position, shape, or composition). With regard to evolutionary and systematic studies, the editors see phylogenetic homology as the only relevant form (e.g., see Chapter 9; Patterson, 1982; and de Pinna, 1991). See Phylogenetic homology, Iterative homology, Biological homology.

Homoplastic. The adjectival form of the term homoplasy. Some authors use homoplasic instead of homoplastic.

Homoplasy. Incongruent data which, on grounds of parsimony, cannot be explained as homology (i.e., as due to common ancestry) but which must be explained by other causes. A number of authors (e.g., Simpson, 1961) have attempted to distinguish between different kinds of homoplasy (e.g., parallelism or convergence) based on presumed recency of common ancestry, adaptation and other hypothetical factors, but we believe that such a distinction cannot be based on empirical grounds.

Hypothesis. A premise assumed to explain certain phenomena without already being accepted as true. Usually considered less encompassing than a theory. A theory is based on hypotheses.

Immunological data. Measures of the degree of reactivity between antibodies and antigens from different species. Comparative immunological methods have been used for phylogenetic work since Nuttall (1904) and are discussed in detail in Maxson and Maxson (1990).

Impetus theory. The medieval theory that states that all movement of extended matter is due to action and reaction between physical bodies. It was important in scholastic philosophy, which viewed God as the first unmoved mover of the universe.

Incertae sedis. A term used to indicate that a taxon cannot be classified with confidence at a particular level. For example, classifying a taxon as "Order: Clupeiformes; Family: *incertae sedis*" would indicate that a taxon cannot confidently be classified to family within the order Clupeiformes. This term is used most often for fossil taxa, frequently because of lack of information (e.g., incomplete preservation).

Incongruence. Homoplasy, or characters that do not show congruence with other data. See Congruence.

Ingroup. The taxonomic group being studied or analyzed phylogenetically in a particular investigation and all the subgroups within it.

Iterative homology. Sometimes referred to as serial homology, homonomy, or homotypy, this type of homology refers to different parts of the same individual organism or organ (e.g., different foliage leaves on a branch, different segments of an individual worm, different limbs of an individual tetrapod).

Jackknifing. A numerical method of resampling (Quenouille, 1956; Efron, 1982) adapted by Lanyon (1985) to calculate indices of support for branching points (nodes) of cladograms or phenograms. Simulated data sets are generated by removing part of the data set (usually a taxon with all of its character information). The resulting data set is smaller than the original data set, and thus jackknifing is not the same as bootstrapping. Although jackknifing could theoretically be used as a resampling technique for taxa or for characters, today it is normally used exclusively for resampling of taxa. Bootstrapping is the resampling technique more often applied to characters. For additional reading see Lanyon (1985, 1987), Felsenstein (1985), and Hillis and Moritz (1990). For a recent criticism of this method see Kluge and Wolf (1993). See Bootstrapping.

Law. An event or sequence of events that always occurs in a predictable way under certain conditions, and hence can be described in mathematical terms. In classical physics, a natural law would specify eternal and immutable causal relations that allow the repeated prediction of the occurrence of events under certain conditions (all else being equal).

Lineage. See Genealogy.

Linnean classification system. A convention of taxonomic ranking that allows representation of taxa as sets and subsets. Major categories are (in descending order of rank) kingdom, phylum, class, order, family, genus, and species. Species names are binomial (two-word) names, and all higher categories are single-word names. For example, the herring species *Clupea harengus* belongs in the genus *Clupea*, the family Clupeidae, the order Clupeiformes, the class Osteichthyes, the phylum Chordata, and the kingdom Animalia. Only the species and genus level names are italicized or otherwise set apart in the typescript. The official rules for construction of family level names and below are published (e.g., Ride et al., 1985, for animals). Additional rank categories between the seven major ranks are available through the use of prefixes such as "sub", "super", "supra", "infra". The rank of tribe is occasionally used as a suprageneric rank, and the word "section" has been used at several different levels.

Macroevolution. Generally used to refer to large-scale evolutionary process at the species level and above.

Macroevolutionary process. See Macroevolution.

Macromutations. A theory of genetic change involving "systemic" or large-scale mutations causing sudden and drastic change with no intermediates between the ancestral and descendant phenotype.

Mapping characters. Placing complex characters (e.g., functional, behavioral) on pre-existing cladograms to analyze them phylogenetically.

Maximum parsimony. A method of cladogram construction that uses the least amount of evolutionary change to explain the data.

Mendelian rules of inheritance. A way of looking at genes as particles, which enabled prediction of certain phenotypic ratios in succeeding generations. Includes "Mendel's laws" (*segregation* of alleles and *independent assortment* of alleles). For further reading, see Ayala and Kiger (1980). See Phenotype.

Metatheory. A theory about another theory or theories.

Microevolution. Generally used to refer to small-scale evolutionary processes working at the level of interbreeding groups of organisms (populations). See Darwinism, Gradualism.

Microevolutionary process. See Microevolution.

Missing data (in phylogenetic analysis). Data unknown for organisms under investigation. Missing data may be the result of incomplete preservation of fossils, or they may result from the evolutionary transformation of a character that renders it incomparable to any other characters known (e.g., particular limb characters in snakes when they are included in analyses with other tetrapods). See Gaps.

Molecular clocks. The molecular clock hypothesis assumes that certain molecules have a constant rate of evolution and therefore that the differences between orthologous molecules of two different species can be used to estimate the age of their closest common ancestor. This notion is strongly supported by some (e.g., Kimura, 1983; Woese, 1987, p. 181; Sibley and Ahlquist, 1987, p. 99) and strongly opposed by others (Goodman et al., 1987, p. 164). For further reading, see Patterson (1987a) and Hillis and Moritz (1990). See Orthology.

Molecular data. Data derived from biochemical components of an organism, such as nucleic acids, amino acids, and proteins. There are many volumes published on molecular systematics, and we recommend Hillis and Moritz (1990) and Patterson (1987). Also see Novacek (Chapter 5) for additional references.

Monophyletic group. A taxon or group of organisms that includes all known descendants of a hypothetical ancestor and no other members. Putatively monophyletic groups are identified by hierarchies of special similarities (also referred to as characters, phylogenetic homologies, or synapomorphies) such as hair and mammary glands for mammals, feathers for birds, bony jaws for gnathostomes, and the presence of bone for osteichthyans.

Monophyletic taxa. See Monophyletic group (monophyletic taxa = monophyletic groups)

Morphological data. Descriptive data based on the anatomy of organisms.

Multistate character. See Character.

Natural groups (of organisms). Monophyletic taxa. See Monophyly, Taxa.

Natural selection. Darwin proposed the theory of natural selection to explain evolutionary change in an evolving population. Under the premises that demographic growth (superfecundity) characterizes every bisexually reproducing population and that natural resources are always limited, the notion of a struggle for existence follows as a logical corollary. Under the premises that a struggle for existence prevails in nature and that offspring in bisexually reproducing organisms are variable, the idea of natural selection must follow as a logical corollary. In other words, variable offspring must compete for limited resources, with some variants showing greater "fitness" than others. Natural selection is measured by the differential survival of the variants. Darwinian fitness is measured in terms of relative reproductive success. For a critical view of the theory of natural selection, see Brady (1982).

Negative evidence. See Cladistics: "negative character information" .

Neodarwinism. A school of evolutionary thought that developed from Darwinism. Variation of offspring in sexually reproducing organisms was a crucial component of Darwin's theory of natural selection (see Natural selection). Darwin's theory of variation proved unacceptable after the rediscovery of Mendel's laws of inheritance at the turn of the century and the subsequent advent of genetics. Neodarwinism denotes the wedding of Darwin's theory of natural selection to the science of genetics. See Darwinism.

Neuroanatomy. Anatomy of the nervous system.

Newtonian ideal. The aim of science to explain observed regularity of phenomena in terms of as-few-as-possible different mechanisms, and preferably a single cause that conforms to the status of a natural law. See Law. Also see Chapters 3 and 9.

Node. A branch point on a cladogram.

"Noise". In phylogenetic analysis, the "noise" is the data that is incongruent with the most parsimonious summary of a data set, explained as homoplasy.

Non-monophyletic group. A number of authors have divided non-monophyletic groups into two categories: paraphyletic and polyphyletic. There is a large body of literature that attempts to clarify or define the distinction between paraphyly and polyphyly (e.g., Hennig, 1966; Nelson, 1971; Farris, 1974; Platnick, 1977; Wiley, 1981; Oosterbroek, 1987; and other references cited in Oosterbroek, 1987). The precise distinction between paraphyly and polyphyly is complex and controversial. Gaffney (1979) suggests the use of non-monophyletic rather than distinguishing between the two terms at all "because of the inconsistencies in simple definitions and the complexity of consistent definitions". Similarly Eldredge and Cracraft (1980, p. 211) note that "a distinction more refined than that between monophyly and non-monophyly hardly seems important". See Monophyletic group.

Numerical taxonomy. Although this term has been used by some authors to refer specifically to phenetics (Sokal and Sneath, 1963), the more general use of the term refers to any method of systematic analysis (e.g., phenetic or cladistic) that uses numerical techniques (e.g., computer analysis of quantitative characters).

Ontogeny. Developmental change throughout the life of an organism. In sexual organisms, this period of change lasts from the fertilized egg cell to death of the organism. See Developmental biology, Developmental pathway.

Orthologous sequences. See Orthology.

Orthology. A term used in molecular studies of proteins or genes that has a meaning equivalent to that of "phylogenetic homology" in studies of morphology. Orthology refers to similarity of genes or gene products between different organisms due to inheritance from a common ancestor. See Phylogenetic homology, Paralogy. For further reading, see Fitch (1970) and Patterson (1987a).

OTU. Operational taxonomic unit. See Terminal taxa.

Outgroup. Any group outside the taxon being studied. The closest outgroup is the sister group. Outgroups are used in comparative studies to help determine whether certain ingroup character states are primitive or derived (i.e., they are used to polarize character information of the ingroup). See Ingroup, Sister group, Polarity, Character.

Outgroup comparison method. The use of outgroups to determine the polarity of ingroup characters. For example, in a phylogenetic analysis of birds, outgroups (e.g., crocodiles, turtles, and other vertebrates) indicate that the presence of feathers is a derived character or synapomorphy of birds, and that certain other features, like the presence of vertebrae, are primitive or plesiomorphic for birds.

Parallelism. See Homoplasy.

Paralogy. A term used in molecular studies of proteins or genes that has a meaning comparable to that of serial homology or iterative homology in studies of morphology. Paralogous genes and gene products may occur in the same organism. See Orthology, Iterative homology. For further reading, see Fitch (1970) and Patterson (1987a).

Paraphyletic group. A type of non-monophyletic group. See Non-monophyletic group.

Parsimony. In systematics, this term refers to the maximum amount of congruence among data. The most parsimonious hypothesis is the one that requires the fewest assumptions (steps of character transformation) about a data set. Parsimony is a necessary methodological tool to select the preferred hypothesis of relationship from a potentially infinite number of possible phylogenetic hypotheses on the basis of character congruence. As Wiley states (1975, p. 236), "only parsimonious hypotheses can be defended by the investigator without resorting to authoritarianism or apriorism". See Congruence.

Pattern. The descriptive order of nature, or, in the words of Eldredge and Cracraft (1980, p. 1), "the apparent orderliness of life", based on regularity of character distribution.

Pattern cladistics. A philosophy that attempts to keep cladistic analysis as independent from evolutionary explanation as possible. In the words of Patterson (1988, p. 72), "If the causal explanation of pattern is to be convincing and efficient, the pattern is better not perceived in terms of explanatory process". This is the argument behind pattern cladistics. See Phylogenetic cladistics.

Pattern reconstruction. Discovery of systematic patterns through analysis of descriptive data.

PAUP. Phylogenetic Analysis Using Parsimony. A computer program developed by Swofford to analyze comparative data to produce the most parsimonious pattern or tree. The most recent version at the time this manuscript was prepared was version 3.1.1 (Swofford and Begle, 1993).

Phenetics. A method of analysis that groups things or organisms according to overall similarity (synapomorphy *and* symplesiomorphy) rather than according to special similarity (synapomorphy only).

Phenogram. A type of branching diagram generated through the use of phenetic methods. The phenogram links taxa through estimates of overall similarity, as opposed to cladograms, which use only putatively derived characters. Phenetics is a method often used to create "keys" for identification purposes (e.g., a key to the birds of Illinois or the fishes of Lake Michigan), but not used much today for investigating phylogeny. See Phenetics.

Phenotype. The observable attributes of an organism. See Genotype.

Phyletic gradualism. See Gradual evolution.

Phylogenetic analysis. Operationally equivalent to the terms cladistics and cladistic analysis. This term should not be confused with Phylogenetic cladistics. See Cladistics, Phylogenetic cladistics.

Phylogenetic cladistics. A philosophy of cladistic analysis that differs from the so-called "pattern cladistics" philosophy in that it accepts evolution as an axiom from which it deduces methods and concepts of systematic research. For further reading see de Queiroz (1988). The differences between pattern cladistics and phylogenetic cladistics are largely philosophical rather than operational. See Pattern cladistics.

Phylogenetic homology. Sometimes referred to as Evolutionary homology. Similarity of structures between different organisms due to inheritance from a common ancestor. In other words, phylogenetic homology is equivalent to synapomorphy. The editors see this as the only form of homology relevant to phylogenetic analysis, although some authors in this volume and elsewhere may not agree.

Phylogenetic tree. See Tree.

Phylogeny. A hypothesis of evolutionary relationships among the members of a monophyletic group. A phylogeny may be fully resolved (normally containing only dichotomous branching) or remain partly unresolved (containing polychotomous branch points or nodes).

Phylogram. A cladogram whose branches are drawn proportional to the amount of proposed character change.

Plesiomorphy or **Plesiomorphic character.** Often referred to as a "primitive" character, a plesiomorphic character is one that is used at a wrong level of inclusiveness. For example, the presence of feathers is a derived character (a synapomorphy) for the group Aves (all birds), but the presence of feathers is a primitive character (a plesiomorphy) for the Aves subgroup Columbidae (the pigeon family). Plesiomorphy is a relative term only. Plesiomorphic characters are never unique to a monophyletic group (they are always more inclusive).

Polarity. In phylogenetic analysis, this is the directional component of character evolution. For example, given two character states within monophyletic group A (e.g., the presence of eyes in subgroup A1 and the absence of eyes in subgroup A2), we can look to the sister group (group B) and other outgroups to polarize this couplet of characters (e.g., if the close outgroups have eyes, then the absence of eyes is the apomorphic state and eyes were secondarily lost within group A). The notions of recapitulation and ontogeny are also sometimes used to polarize characters by considering early embryonic states to be indicative of phylogenetically primitive character states. Biostratigraphy has also been used to influence theories of phylogeny, and thereby indirectly polarize characters. See Sister group, Outgroup comparison method, Apomorphy, Recapitulation, Biostratigraphy.

Polyphyletic group. See Non-monophyletic group.

Population. A group of interbreeding organisms in the case of bisexually reproducing organisms. In asexual organisms, a population is a locally distributed clone.

Positive evidence. See Cladistics.

Process. In evolutionary studies, process is the causal explanation for observed order or patterns in nature.

Punctuated equilibrium. The theory that morphological and genetic changes tend to occur during a relatively rapid process of speciation, whereas the genotype and phenotype remain more or less stable during the relatively long duration of the species. The equilibrium characterizing a species' genotype and phenotype is "punctuated" during speciation. This supposedly explains some of the morphological "gaps" in the fossil record. The suggestion has been made (without much empirical evidence) that the duration of a speciation process is about 10% of the total "lifetime" of a species. This theory is discussed at length in Eldredge and Gould (1972) and Gould and Eldredge (1977). See Gaps (in morphology).

Punctuationism. A general term referring to theoretical scenarios of discontinuous species change (e.g., punctuated equilibrium, quantum evolution). See Punctuated equilibrium, Quantum evolution.

Putative. Hypothesized or presumed.

Qualitative. Having to do with characteristic (generally non-numerical) elements, as opposed to quantitative. See Quantitative.

Quantitative. Numerical. A quantitative method involves mathematical equations and produces exact numerical solutions, as opposed to qualitative. See Qualitative.

Quantum evolution. A term proposed by Simpson (1944) for rapid evolutionary transformation. Quantum speciation was a term used by Stebbins (1977) to refer to punctuational modes of speciation in plants (by the development of polyploidy for example). See also Punctuated equilibrium.

Range data (in historical biogeography). The total geographic area inhabited by a taxon.

Rational morphology. The name given to a theory that seeks to explain morphology on the basis of structural laws inherent in organic matter (rather than by extraneous forces shaping organic matter, e.g., natural selection). See Law, Structuralism. For further reading see Chapters 7 and 9.

Recapitulation. Refers to the notion that ontogeny (the embryonic development of an organism) recapitulates or repeats phylogeny (the evolutionary history of the lineage including that organism). The ontogenetic transformations of a character could therefore reveal the evolutionary history of that character through recapitulation. See also Biogenetic law, Haeckel's Law. For further reading, see Løvtrup (1978).

Reciprocal illumination. Refers to the idea that scientific knowledge increases by continuous feedback between observation and explanation.

Reproductive isolation. When two populations are separated from each other so they can no longer interbreed with each other.

Reticulate speciation. A process theory involving the origin of a new species through hybridization of two different species.

Selection. See Natural selection.

Serial homology. See Iterative homology.

Sister group. The closest monophyletic group outside the ingroup. See Ingroup, Monophyletic group.

Speciation. The process leading to the origin of new species through time. See Anagenesis, Cladogenesis, Reticulate speciation.

Species. There probably are as many species definitions as there have been authors defining species. There are, however, a number of attributes that most definitions have in common:

(1) Reproductive coherence: due to genetic and behavioral compatibility of the sexes (in the case of sexually reproducing organisms).
(2) Uniqueness of its evolutionary role: due to the genetic isolation from other species.
(3) A beginning in time: due to speciation.
(4) An end in time: due to cladogenesis, terminal extinction (predicted also for extant species).

These attributes are generally applied by systematists who believe that species are real units in nature (e.g., Eldredge and Cracraft, 1980, p. 15) that are somehow different from higher taxonomic groups, which are simply groups of species. But there are also systematists who believe that species are no more or no less real than higher taxa, and that they should be defined simply as the smallest identifiable monophyletic group. The species definition problem is a controversial one for many reasons, and we cannot begin to address all of the issues involved here. The issue of species being "real" or "individuals" in nature is discussed further in Ghiselin (1974) and Hull (1976, 1978). The problems and controversies surrounding species have little or no effect on phylogenetic analysis, because cladistic methods are concerned only with monophyletic groups, regardless of what one chooses to call those groups. See Phylogenetic analysis, Cladistics.

Spreading ridge. The mid-oceanic borders between major plates of the earth's surface, where the plates are moving away from each other and new crust is being formed. Part of the theory of plate tectonics (and drifting continents).

Stratigraphy. The study of layered rocks, particularly with regard to their sequence in time and correlation of similar sequences from different areas.

Stratocladistics. A theory that allows the stratigraphic distribution of fossils to influence the evaluation and choice of possible trees. For further reading, see Fisher (Chapter 6).

Strict consensus tree. A type of consensus tree that summarizes only the branching information repeated in all of the original trees. See Consensus tree.

Structuralism. A school of science that defends the theory of "rational morphology" which results in an ahistorical theory of (organismic) form. See Rational morphology.

Structuralist. See Structuralism.

Subclade. Clade within a larger clade. See Clade, Monophyletic group.

Synapomorphy. A shared derived character or character state. A character diagnosing a monophyletic group (e.g., feathers for birds). See Monophyletic group, Characters.

Synthetic theory of organic evolution. The effort to relate neodarwinism (the combination of modern genetics with Darwin's theory of natural selection) to other domains of natural history (paleontology, biogeography, ecology, and behavior) to provide for a universal explanation of evolutionary processes. See Neodarwinism.

Systematic pattern. See Cladogram.

Systematics. A field of study that seeks to identify, describe, and explain the orderliness of nature.

Taphonomy. The study of the process of fossilization and what happens to organisms after death.

Taxa. Plural form of taxon. See Taxon.

Taxic approach. The taxic approach (Eldredge, 1979) is one that uses the phylogenetic analysis of taxa (cladograms), rather than hypothetical transformation series (e.g., stratigraphy or ontogeny), to construct hypotheses of evolutionary history.

Taxon. A group of organisms. A taxon usually has a proper name, and can be (and should be, in our opinion) monophyletic (e.g., the genus *Clupea*, which includes the true herrings, or the family Acipenseridae, which includes the sturgeons). Occasionally taxon is also used to refer to non-monophyletic groups such as the traditional Reptilia (a group that excludes some of its putative descendants, i.e., birds) or species as defined by various authors (see discussion in Chapter 9). Only monophyletic groups are considered to be natural taxa here. We agree with Nelson and Patterson (1993, p. 442) that "taxa are discovered through characters that are themselves discovered through empirical investigation". See Character, Monophyletic group.

Taxonomic congruence. Congruence between different cladograms when each cladogram is based on a different data set. For example, if a study of skeletal characters indicates A and B are more closely related to each other than to C, and another study based on muscle characters or molecular data indicates A and B are more closely related to each other than to C, then there is taxonomic congruence between these two studies.

Taxonomic group. See Taxon. (taxonomic group = a taxon)

Teleology. A philosophy that stipulates that everything in nature has a predetermined purpose. See also discussion by Mayr (1982).

Temporal patterns. Patterns through time based on stratigraphic data. See Stratigraphy.

Terminal taxa. The taxa used in a cladogram, or the distal tips of a phylogenetic tree (e.g., sp. A, sp. B, and sp. C in Chapter 4, Fig. 1). They are usually species, but they can also be higher-level taxa (when monophyletic groups of species are collapsed into single branches) or subspecific groups. A terminal taxon is equivalent to an "OTU" or an operational taxonomic unit (the taxa that provide the data for phylogenetic analysis).

Test. Re-examination of a theory in an attempt to falsify or corroborate previous results. A theory is open to falsification (or corroboration) if it allows predictions that can be put to an empirical test. Most scientists agree that a theory needs to be testable in order to be accepted as scientific theory (in opposition, for example, to beliefs).

Theoretical pluralism. The notion that multiple alternative explanations are required for every set of phenomena (in opposition to the Newtonian ideal). See Beatty (this volume).

Theory. See Hypothesis.

Three-taxon statement. The most fundamental statement in phylogenetic systematics, which specifies that taxon A is more closely related to taxon B than either A or B is to C–or any permutation thereof ((A,B)C), ((A,C)B), (A(B,C)). For example, see Chapter 4, Figure 1.

Topology (with regard to comparative anatomy). Relative position. To use topology to identify homology is to find a structure in the same relative position in two different organisms (e.g., the forearm of a human and the wing of a bird) and consider them homologs on the basis of their position.

Topology (with regard to phylogenetic tree structure). The relative order and position of taxa and branch-points on a phylogenetic tree; usually exclusive of individual branch length.

Total evidence approach. The approach that argues for combining all data into a single set for analysis. Sometimes discussed in opposition to "taxonomic congruence method". See discussions on pp. 76-79 and pp. 119-120. See Taxonomic congruence.

Total evidence information. See Total evidence approach.

Transformed cladists. See Pattern cladists.

Tree (X-tree, phylogenetic tree). In graph theory, the notion of tree encompasses any branching diagram, and hence is equivalent to the notion of cladogram (Hendy and Penny, 1984). An X-tree results if the time dimension is superimposed on the cladogram (on the systematic pattern of character distribution), and the nodes in the cladogram are taken as hypothetical common ancestors of the diverging clades (Patterson, 1983). A phylogenetic tree corresponds to a branching diagram that is postulated to represent the actual historical course of phylogeny, specifying particular ancestors and descendants.

Tree-length. Tree-length gives the total number of character state changes in a tree (or cladogram). Every character (or character state) change implied by the tree (from plesiomorphic to apomorphic character states, as well as convergence and reversals) is counted as one evolutionary step, and the number of evolutionary steps is then added for the whole tree (or cladogram).

Unconventional characters. Characters belonging to systems not usually used for phylogenetic analyses (e.g., sperm morphology). Referred to as "non-traditional" characters by some authors (e.g., Wake, 1993). Unconventional characters can become conventional characters if their use increases sufficiently. For example, at one time molecular characters were unconventional, but with the large increase in molecular data over the last ten years, they have become conventional. See Chapter 7.

Vicariance. A name for the process that occurs when a formerly continuous population is divided by the appearance of a barrier.

Weighting characters. Placing a heavier emphasis on some characters than others for phylogenetic analysis.

Whig history. A misrepresentation of history or a historical event (often by taking it out of context) in such a way that it supports a present-day argument, theory, or party line. For further reading, see Mayr (1982).

Acknowledgments

We thank Scott Lanyon, Rudiger Bieler, Petra Sierwald, and Mario de Pinna for reading and commenting on this list of terms.

References

Adams, E. N. 1972. Consensus techniques and the comparison of taxonomic trees. *Systematic Zoology*, 21(4): 390-397.

Ayala, F. J., and J. A. Kiger, Jr. 1980. *Modern Genetics*. Menlo Park, CA: Benjamin/Cummings Publishing Co. Inc. 844 pp.

Barrett, M., M. J. Donoghue and E. Sober. 1991. Against consensus. *Systematic Zoology*, 40(4): 486-493.

Brady, R. 1980. Natural selection: an examination of the criteria by which a theory is judged. *Systematic Zoology*, 28(4): 600-621.

Brady, R. 1982. Dogma and doubt. *Biological Journal of the Linnean Society*, 17: 79-96.

Brown, J. H., and A. C. Gibson. 1983. *Biogeography*. St. Louis, MO: C. V. Mosby Co.

Brundin, L. Z. 1981. Croizat's panbiogeography versus phylogenetic biogeography. In G. Nelson and D. E. Rosen (Eds.), *Vicariance Biogeography: A Critique*, 94-138. New York: Columbia

University Press. 593 pp.
de Beer, G. 1958. *Embryos and ancestors*. Oxford: Oxford University Press.
de Pinna, M. C. C. 1991. Concepts and tests of homology in the cladistic paradigm. *Cladistics*, 7: 367-394.
de Queiroz, K. 1988. Systematics and the Darwinian revolution. *Philosophy of Science*, 55: 238-259.
Dodds, K. G. 1986. Resampling methods in genetics and the effect of family structure in genetic data. Institute of Statistics Mimeo. Series 1684T. Raleigh: North Carolina State University.
Efron, B. 1979. Bootstrap methods: another look at the jackknife. *Ann. Statist.* 7: 1-26.
Efron, B. 1982. The jackknife, the bootstrap, and other resampling plans. CBMS-NSF regional conference series in applied mathematics No. 38. Philadelphia, PA: Society for Industrial and Applied Mathematics.
Eldredge, N. 1979. Alternative approaches to evolutionary theory. *Bulletin of the Carnegie Museum of Natural History*, 13: 7-19.
Eldredge, N., and J. Cracraft. 1980. *Phylogenetic Patterns and the Evolutionary Process*. New York: Columbia University Press. 349 pp.
Eldredge, N., and S. J. Gould. 1972. Punctuated equilibria: an alternative to phyletic gradualism. In T. J. M. Schopf (Ed.), *Models in Paleobiology*, 82-115. San Francisco: Freeman, Cooper and Co.
Farris, J. S. 1974. Formal definitions of paraphyly and polyphyly. *Systematic Zoology*, 23: 548-554.
Farris, J. S. 1988. *Hennig86 version 1.5 manual*, software and MSDOS program.
Felsenstein, J. 1985. Confidence limits on phylogenies: an approach using the bootstrap. *Evolution*, 39: 783-791.
Fitch, W. M. 1970. Distinguishing homologous from analogous proteins. *Systematic Zoology*, 27: 27-33.
Forey, P. L., C. J. Humphries, I. J. Kitching, R. W. Scotland, D. J. Siebert and D. M. Williams. 1992. *Cladistics: A Practical Course*

in Systematics. The Systematics Association publication number 10. Oxford: Oxford Press. 191 pp.

Futuyma, D. J. 1979. *Evolutionary Biology*. Sunderland, MA: Sinauer Associates Inc. 565 pp.

Gaffney, E. S. 1979. An introduction to the logic of phylogeny reconstruction. In J. Cracraft and N. Eldredge (Eds.), *Phylogenetic Analysis and Paleontology*, 97-111. New York: Columbia University Press.

Ghiselin, M. T. 1974. A radical solution to the species problem. *Systematic Zoology*, 23: 536-544.

Goodman, M., M. M. Miyamoto and J. Czelusniak. 1987. Pattern and process in vertebrate phylogeny revealed by coevolution of molecules and morphologies. In C. Patterson (Ed.), *Molecules and Morphology in Evolution: Conflict or Compromise?* pp. 141-176. Cambridge: Cambridge University Press.

Gould, S. J. 1970. Dollo on Dollo's Law: irreversibility and the status of evolutionary laws. *Journal of the History of Biology*, 3: 189-212.

Gould, S. J., and N. Eldredge. 1977. Punctuated equilibria: the tempo and mode of evolution reconsidered. *Paleobiology*, 3: 115-151.

Hendy, M. D., and D. Penny. 1982. Branch and bound algorithms to determine minimal evolutionary trees. *Mathematical Biosciences*, 59: 277-290.

Hendy, M. D., and D. Penny. 1984. Cladograms should be called trees. *Systematic Zoology*, 33: 245-247.

Hennig, W. 1966. *Phylogenetic Systematics*. Urbana: University of Illinois Press. 263 pp.

Hillis, D. M., and C. Moritz. 1990. An overview of applications of molecular systematics. In D. M. Hillis and C. Moritz (Eds.), *Molecular Systematics*, 502-515. Sunderland, MA: Sinauer Associates Inc.

Holder, N. 1983. Developmental constraints and the evolution of vertebrate digit patterns. *Journal of Theoretical Biology*, 104: 451-471.

Hull, D. L. 1976. Are species really individuals. *Systematic Zoology*, 25: 174-191.

Hull, D. L. 1978. A matter of individuality. *Philosophy of Science*, 45: 335-360.

Humphries, C. J., and L. Parenti. 1986. Cladistic biogeography. *Oxford Monographs in Biogeography*, 2: 1-98.

Kimura, M. 1983. *The Neutral Theory of Molecular Evolution*. Cambridge: Cambridge University Press.

Kitching, I. J. 1992. Tree-building techniques. In P. L. Forey, C. J. Humphries, I. J. Kitching, R. W. Scotland, D. J. Siebert and D. M. Williams (Eds.), *Cladistics*, 44-66. Systematics Association publication no. 10. Oxford: Clarendon Press.

Kluge, A. G., and A. J. Wolf. 1993. Cladistics: what's in a word? *Cladistics*, 9: 183-199.

Lanyon, S. M. 1985. Molecular perspective on higher-level relationships in the Tyrannoidea (Aves). *Systematic Zoology*, 34: 404-418.

Lanyon, S. M. 1987. Jackknifing and bootstrapping: important "new" statistical techniques for ornithologists. *Auk*, 104: 144-146.

Løvtrup, S. 1978. On von Baerian and Haeckelian recapitulation. *Systematic Zoology*, 27: 348-352.

Maxson, L. R., and R. D. Maxson. 1990. Proteins II: immunological techniques. In D. M. Hillis and C. Moritz (Eds.), *Molecular Systematics*, 127-155. Sunderland, MA: Sinauer Associates Inc.

Mayr, E. 1982. *The Growth of Biological Thought*, 974 pp. Cambridge, MA: Harvard University Press.

Murphy, R. W., J. W. Sites, Jr., D. G. Buth and C. H. Haufler. 1990. Proteins I: isozyme electrophoresis. In D. M. Hillis and C. Moritz (Eds.), *Molecular Systematics*, 45-126. Sunderland, MA: Sinauer Associates Inc.

Nei, M. 1991. Relative efficiencies of different tree-making methods for molecular data. In M. M. Miyamoto and J. Cracraft (Eds.), *Phylogenetic Analysis of DNA Sequences*, 90-128. New York, Oxford: Oxford University Press.

Nelson, G. J. 1971. Paraphyly and polyphyly: redefinitions. *Systematic Zoology*, 20: 471-472.

Nelson, G. 1978a. Ontogeny, phylogeny, paleontology and the

biogenetic law. *Systematic Zoology*, 27: 324-345.

Nelson, G. 1978b. From Candolle to Croizat: comments on the history of biogeography. *Journal of the History of Biology*, 11(2): 269-305.

Nelson, G. 1993. Why crusade against consensus? a reply to Barrett, Donoghue, and Sober. *Systematic Biology*, 42(2): 215-216.

Nelson, G., and C. Patterson. 1993. Cladistics, sociology and success: a comment on Donoghue's critique of David Hull. *Biology and Philosophy*, 8: 441-443.

Nelson, G., and N. Platnick. 1981. *Systematics and Biogeography: Cladistics and Vicariance*. New York: Columbia University Press. 567 pp.

Nelson, G., and D. E. Rosen (Eds.). 1981. *Vicariance Biogeography: A Critique*. New York: Columbia University Press. 593 pp.

Nuttall, G. H. F. 1904. *Blood Immunity and Blood Relationship*. Cambridge: Cambridge University Press.

Oosterbroek, P. 1987. More appropriate definitions of paraphyly and polyphyly, with a comment on the Farris 1974 model. *Systematic Zoology*, 36(2): 103-108.

Patterson, C. 1978. *Evolution*. London: British Museum (Natural History). 197 pp.

Patterson, C. 1982. Morphological characters and homology. In K. A. Joysey and A. E. Friday (Eds.), *Problems in Phylogenetic Reconstruction*, 21-74. London: Academic Press.

Patterson, C. 1983. How does phylogeny differ from ontogeny. In B. C. Goodwin, N. Holder and C. C. Wylie (Eds.), *Development and Evolution*, 1-31. Cambridge, U.K.: Cambridge University Press.

Patterson, C. 1987a. Introduction. In C. Patterson (Ed.), *Molecules and Morphology in Evolution: Conflict or Compromise?*, 1-22. Cambridge: Cambridge University Press.

Patterson, C. (Ed.). 1987b. *Molecules and Morphology in Evolution: Conflict or Compromise?*, 1-229 pp. Cambridge: Cambridge University Press.

Patterson, C. 1988. The impact of evolutionary theories on systematics. In D. L. Hawksworth (Ed.), *Prospects in Systematics*, 59-91. Oxford: Clarendon Press.

Pauper, K. 1976. *Unended Quest*. Illinois: Open Court Publishing Company. 255 pp.

Platnick, N. I. 1977. Paraphyletic and polyphyletic groups. *Systematic Zoology*, 26: 195-200.

Quenouille, M. 1956. Notes on bias in estimation. *Biometrika* 43: 353-360.

Ride, W. D. L., C. W. Sabrosky, G. Bernardi, R. V. Melville, J. O. Corliss, J. Forest, K. H. L. Key, C. W. Wright. 1985. *International Code of Zoological Nomenclature*. 3rd edition. Berkeley: University of California Press. 338 pp.

Roth, V. L. 1984. On homology. *Biological Journal of the Linnean Society*, 22: 13-29.

Sanderson, M. J. 1989. Confidence limits on phylogenies: the bootstrap revisited. *Cladistics*, 5: 113-129.

Schoch, R. M. 1986. *Phylogeny Reconstruction in Paleontology*. New York: Van Nostrand Reinhold.

Simpson, G. G. 1944. *Tempo and Mode in Evolution*. New York and London: Hafner Publishing Co. 237 pp.

Sokal, R. R., and P. H. Sneath. 1963. *Principles of Numerical Taxonomy*. San Francisco: Freeman Publishing Co.

Stebbins, G. L. 1977. *Processes of Organic Evolution*, 3rd ed. Englewood Cliffs, NJ: Prentice-Hall Inc.

Swofford, D. L., and D. P. Begle. 1993. *PAUP: Phylogenetic Analysis Using Parsimony, Version 3.1*. [Manual version is 3.1, updated floppy disc is version 3.1.1]. Washington, D.C.: Laboratory of Molecular Systematics, Smithsonian Institution. 257 pp.

Swofford, D. L., and G. J. Olsen. 1990. Phylogeny reconstruction. In D. M. Hillis and C. Moritz (Eds.), *Molecular Systematics*, 411-501. Sunderland, MA: Sinauer Associates Inc.

Wagner, G. P. 1989. The biological homology concept. *Annual Review of Ecology and Systematics*, 20: 51-69.

Wake, M. H. 1993. Non-traditional characters in the assessment of caecilian phylogenetic relationships. *Herpetologica* Monographs, 7: 42-55.

Wiley, E. O. 1975. Karl R. Popper, systematics, and classification: a reply to Walter Bock and other evolutionary taxonomists. *Systematic Zoology*, 24: 233-243.

Wiley, E. O. 1981. *Phylogenetics. The Theory and Practice of Phylogenetic Systematics*. New York: J. Wiley and Sons. 439 pp.

Wiley, E. O., D. J. Siegel-Causey, D. R. Brooks and V. A. Funk. 1991. *The Compleat Cladist: A Primer of Phylogenetic Procedures*. Lawrence: Museum of Natural History, University of Kansas. 158 pp.

List of Contributors

Beatty, J.	Department of Ecology, Evolution and Behavior, University of Minnesota, Minneapolis, MN 55455
Brady, R. H.	Philosophy, Ramapo College, Mahwah, NJ 07430
Fisher, D. C.	Museum of Paleontology and Department of Geological Sciences, University of Michigan, Ann Arbor, MI 48109
Grande, L.	Department of Geology, Field Museum of Natural History, Roosevelt Road at Lake Shore Drive, Chicago, IL 60605
Novacek, M. J.	Office of the Vice-President, and Dean of Science, American Museum of Natural History, Central Park West at 79th Street, New York, NY 10024
Rieppel, O.	Department of Geology, Field Museum of Natural History, Roosevelt Road at Lake Shore Drive, Chicago, IL 60605
Shubin, N. H.	Department of Biology, University of Pennsylvania, Philadelphia, PA 19104
Wake, M. H.	Department of Integrative Biology and Museum of Vertebrate Zoology, University of California, Berkeley, CA 94720

Subject Index

Words in **bold** denote terms included in preceding glossary. For many terms, most pages with only casual use of the term are not listed.